Introduction
to the
Renormalization Group
and to
Critical Phenomena

Introduction to the Renormalization Group and to Critical Phenomena

Pierre Pfeuty and Gérard Toulouse

Université de Paris Sud, Centre d'Orsay

Translation:

G. Barton

University of Sussex

A Wiley–Interscience Publication

JOHN WILEY & SONS

London · New York · Sydney · Toronto

First published 1975 © Presses Universitaires de Grenoble under
the title Introduction au Groupe de Renormalisation et à ses
applications by Gérard Toulouse and Pierre Pfeuty

Library of Congress Cataloging in Publication Data:
Toulouse, Gérard.
  Introduction to the renormalization group and to critical
phenomena.

  Translation of Introduction au groupe de renormalisation et à
ses applications.
  Authors' names in reverse order in original French ed.
  'A Wiley–Interscience publication.'
  Includes bibliographical references and index.
  1. Critical phenomena (Physics) 2. Renormalization (Physics)
I. Pfeuty, Pierre, joint author. II. Title.
QC173.4.C74T6813   + 536.401   76-26111

ISBN 0 471 99440 5

Photosetting by Technical Filmsetters Europe Limited, Manchester.
Printed and bound in Great Britain by The Pitman Press, Bath

Contents

Preface

This book is intended as an introduction to a very extensive domain within physics, which has grown by 'horizontal' stratification around certain problems arising in field theory, statistical mechanics, solid-state physics, chemical physics and elsewhere. In the space of a few years we have progressed from a set of rather vague analogies to a domain thoroughly unified by its methodology and by the language of the renormalization group. It would seem that without exaggeration one can detect here a new level of understanding in physics, endowed with an original and powerful conceptual system to which it now becomes necessary to facilitate access by non-specialists.

It is appropriate to picture this domain as a continent under exploration, with some regions partly known (critical phenomena in phase transitions, various problems in field theory, the statistical mechanics of polymers, the Kondo effect, etc.), others hardly touched (dynamical critical phenomena, inhomogeneous media, one-dimensional metals, instabilities in steady states—hydrodynamics, chemistry, biology—and many problems of elementary particles, etc.), and with its confines quite untraced as yet. In this respect one can already speculate about the possibilities of application or inspiration in science as a whole, (and, why not, in social studies).

In such circumstances we wanted to write a truly introductory book, as thin and as little burdened by erudition as possible, and aiming to give a clear outline of the simplest ideas. Hence we have not attempted a detailed discussion such as can nowadays be given of, say, quantum mechanics. In the field in which we are interested, such manuals will undoubtedly be provided in due course, which is to say later. Instead, the present book intends to use examples and illustrations primarily in order to familiarize the reader with certain concepts and with a certain language, so that he should be able to appropriate these and to exploit them in his own field.

There is no doubt that, hitherto, the renormalization-group approach has proved especially fruitful in the theory of critical phenomena in phase transitions. For this as well as for pedagogic reasons, we have chosen to centre the book on the theory of critical phenomena, though, throughout, we stress the generality of the method and of the results, and indicate applications to other problems where appropriate.

The first three chapters provide a wholly elementary introduction to the theory of critical phenomena, intended to make the book accessible to any reader having a basic familiarity with science. The slant towards the theory of critical phenomena allows the vocabulary and the results of the renormalization-group approach to be introduced naturally, and provides convenient access to a point of view which it is more awkward to reach in other ways; precisely the reason why the renormalization group remained for so long the exclusive preserve of a small set of theorists working with relativistic fields. Even when similar methods were exploited in quantum statistical mechanics (for the Kondo effect, i.e. for the problem of dilute magnetic impurities in metals), the circle of initiates hardly widened; and everyone agrees that it was through the theory of critical phenomena that the full power of the method was actually revealed, put to work, and generalized.

This book is designed for students of physics (at first-year postgraduate or advanced undergraduate level), for experimental or theoretical physicists, and also for the community of mathematicians, hydrodynamicists, chemists, biologists and others, who should not be bypassed by the rapid development of a new theoretical approach, pioneered admittedly by physicists, but which by its generality bridges the divisions between several disciplines.

The renormalization-group approach leads to a topological description of the phenomena, in an abstract space, namely in parameter space; more precisely, it leads to differential equations whose solutions specify trajectories in this space. From this point of view the method is conceptually akin to the mathematical theory of dynamical systems, which has already been applied to problems as diverse as the competition between species in ecology, the course of chemical reactions, biological clocks and morphogenesis. By reason of their global approach and of their insistence on universality properties, these theories constitute a whole new way of thinking, and it seems reasonable to locate renormalization-group theory in this larger context.

Chapter 1 introduces the most basic ideas about phase transitions: the order parameter, broken symmetry, fluctuations, correlation length, scaling laws, homogeneity properties, and also the role of the dimensionalities and the concept of universality. Chapters 2 and 3 elaborate these ideas at greater length while describing the approaches predating the renormalization group. Chapter 1 also introduces the main concepts of the renormalization-group approach: the systematic reduction in the number of degrees of freedom, covariance under dilatation, parameter space, critical surface, fixed points, relevance, stability, competition and crossover. Our aim is to explain at the outset how the renormalization-group approach leads, very suggestively, to a geometric and topological description of critical phenomena. The more quantitative development of these ideas begins in Chapter 4, and their implementation for dimensionalities $d = 4 - \varepsilon$ is presented in Chapter 5. One of the merits of the renormalization group as applied to critical phenomena is that it admits of a reasoned classification of these; thus, one can distinguish between simple and complex systems, rather like chemistry distinguishes between elements and compounds. Chapter

6 presents the results obtained for simple systems. Chapter 7 introduces the study of complex systems. Chapters 8 and 9 discuss two examples of perturbations that are especially important both for their intrinsic physical interest, and as paradigms. Chapter 10 summarizes the results on perturbations by constant fields, while Chapter 11 deals with the effects of couplings to other degrees of freedom. Tricritical (and polycritical) points are studied in Chapter 12. Given the idea of marginality, local or persistent, one can consider various pathological situations. Chapter 13 shows how the renormalization-group approach allows one to understand the exceptions as well as the rules, and how it yields valuable analogies between quite different physical systems (systems of reduced dimensionality, Kondo effect, one-dimensional metals, etc).

Because our own approach is primarily descriptive, we have been careful to complement it by giving references to aid the reader who wishes to pursue further either calculational methods and their mathematical bases, or comparisons with experimental results. The available articles which review the renormalization-group approach are addressed to an already expert readership, and tend to stress particular aspects of the domain defined above. They should serve as a useful bridge between the present book and the study of the specialized research literature. Such reviews are listed under the references to Chapter 4.

Our attitude to referencing and to attributions may seem cavalier to some readers. The names of authors have been used simply for orientation, and references have been restricted to the bare minimum necessary to unlock the doors to the original literature, which is immense. Authors whose work has not been cited are asked for their forgiveness, and we hope that they will read no malice into our attempt to secure brevity as the price for increased coverage.

The book was first projected when one of the authors (G. T.) gave a first-year postgraduate course in solid-state physics at the Université de Paris-Sud, Centre d'Orsay.

The authors take pleasure in thanking Messieurs A. Blandin, J. Friedel, P. G. de Gennes and P. Lacour-Gayet for their help and encouragement. They are grateful also to Mesdames H. Coulle and A. Touchant for their assistance in preparing the text and the figures.

August 1974

Translator's Note

For the English edition the authors have corrected misprints in equations in the original, and have supplied relatively brief additional paragraphs, chiefly in Sections 2.6, 7.4 and 11.3, and in Appendix 2.2. Apart from these the English text is a straight translation of the French first edition.

May 1976

CHAPTER 1
Introduction

'I too of course play games with symbols...., but in a way that keeps me mindful that I am only playing' Kepler

The structure described in this book rests on twin supports, one consisting of general ideas and the other of specific applications. Each reinforces the other. It is customary in an introductory chapter to concentrate mainly on the general concepts, in order to obtain a rapid overall orientation as to the problems, their interconnections and their proper context in physics as a whole. Admittedly, a bird's eye view imparted so quickly may induce some dizziness; the more concrete developments contained in the later chapters are needed to elucidate the concepts and to establish their applicability. Indeed it is precisely because the specific applications and the general ideas cross-illuminate each other that the subject has now reached a stage where one can understand, fairly quickly, what has been achieved already, and where one can foresee future developments. Hence the reader will do well to refer back to this introduction at appropriate stages later in the book.

1.1 Introduction to phase transitions

Phases, phase transitions, and phase diagrams

Consider water and its various phases: solid (ice), liquid, and gas (steam), all so easily identified in everyday experience. A change in temperature allows one to observe the transitions from one phase to another: thawing (freezing) and boiling (condensation), which are so familiar that they have long served to define the temperature scale (centigrade). There is something fascinating in these phase transitions, particularly in their aspects of qualitative change, discontinuity and transmutation; a small shift in temperature or in some other parameter suffices to trigger a spectacular alteration. Moreover they belong to a very widespread family: there exist phase transitions that are magnetic, ferroelectric, superfluid, superconducting; others entail phase separation in solutions, order–disorder transitions in alloys, and mesomorphic transitions in liquid crystals. It is hardly surprising that the discovery and elucidation of new phases continues to play an important role in the physics of condensed matter. One takes a given system, varies some of its parameters like temperature, pressure, or external field strengths, in order to exhibit the different phases of the system and their domains of existence, and enters the results on a chart called a phase diagram.

1

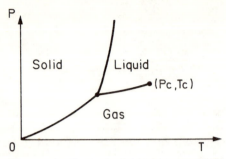

Figure 1.1.   Typical solid–liquid–gas phase
diagram

In order to introduce some ideas we shall consider three particular phase diagrams.

The first (Figure 1.1) is a pressure against temperature diagram showing the domains of existence of three phases, solid, liquid and gas. Note two special points on this diagram: the first, called a triple point, at the junction of three domains; the second, called a critical point, the end point of a line dividing the liquid from the gaseous region. By circling round the critical point (P_c, T_c) one can pass continuously from the liquid to the gaseous phase, without any necessity for a discontinuous transition.

The second example (Figure 1.2) is a magnetic-field against temperature diagram for a material undergoing a transition of the ferromagnetic type; it exhibits a boundary line along the horizontal axis, terminating at a point called the critical point, as in the preceding case. In zero field and at high temperature one observes a disordered phase called the paramagnetic phase, having no magnetization; as the temperature drops, a transition takes place at the critical point $T = T_c$, and for $T < T_c$ one observes an ordered phase called the ferromagnetic phase, which is spontaneously magnetized. There is an analogy between Figure 1.2 and that part of Figure 1.1 which concerns the liquid-to-gas transition; the magnetic field and the pressure play analogous roles. In Figure 1.2, on crossing the boundary $(H = 0, 0 < T < T_c)$, the magnetization changes

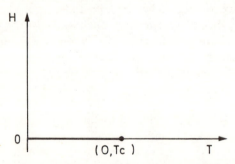

Figure   1.2.   Typical   ferromagnetic   phase
diagram

discontinuously; the discontinuity decreases as one crosses the boundary closer to the critical point; at the critical point the discontinuity vanishes.

The third diagram (Figure 1.3), too, is a magnetic-field against temperature diagram, but now for a material undergoing a transition of the antiferromagnetic type, with the magnetization alternating in direction in the ordered phase. On this diagram one notes two domains, separated by a continuous line part of which is drawn solid and part broken. This corresponds to the following distinction: on crossing the solid line one observes a discontinuous jump in the magnitude of the alternating magnetization; by contrast there is no such discontinuity on crossing the broken line. One says that the broken portion of this transition line is a line of critical points (since there is no discontinuity); the line of critical points has an end point, on the transition line, called a tricritical point.

Order parameter, broken symmetry and the order of a phase transition

How can one understand why there is a succession of different phases as the temperature rises? The reason is that in the free energy $F = U - TS$, which is minimized by the state of thermodynamic equilibrium, there is competition between the energy U (which favours order) and the entropy S (which favours disorder); a rise in temperature tilts the balance towards entropy and disorder. To distinguish between two phases one defines an order parameter, having non-zero value in the ordered phase, and zero value in the disordered phase (which, in general, is the high-temperature phase, though there are exceptions to this rule when different degrees of freedom are coupled). For a ferromagnetic transition, as in Figure 1.2, the order parameter is the homogeneous magnetization; for an antiferromagnetic transition, Figure 1.3, it is the magnitude of the alternating magnetization. In general, a non-zero value of the order parameter corresponds to the breaking of a symmetry. Thus, in our two magnetic examples, the broken symmetry is the symmetry under rotations; in the high-temperature phase the system is invariant under rotations around all three axes, while in the low-temperature phase it remains invariant only under rotations around one axis, namely the axis of spontaneous magnetization. In the case of the liquid-to-gas transition, Figure 1.1, strictly speaking there is no broken symmetry; nevertheless

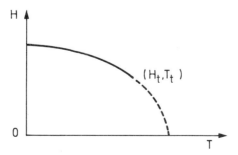

Figure 1.3. Typical antiferromagnetic phase
diagram (metamagnetic case)

4

one defines an order parameter through the difference in density between liquid and gas.

When the order parameter has a discontinuity at the transition, one says, following Landau, that one is dealing with a first-order transition (Landau and Lifshitz, 1958; Stanley, 1971); where there is no such discontinuity, the transition is second order. In Figures 1.1 and 1.2, the transition lines are lines of first-order transitions; at the critical points the transitions are of second order. In Figure 1.3, the transition line has a segment of first order and another segment (the line of critical points) of second order; at the tricritical point, the transition is still of second order according to our definition, there being no discontinuity in the order parameter. If one plots the measured value of the order parameter as a function of temperature for a second-order transition, one obtains typically a curve like the one shown in Figure 1.4: there is no discontinuity in the order parameter itself, but there is a discontinuity in its slope.

In view of all this, any programme for investigating phase transitions will be concerned with the problems of specifying the order parameter, predicting the order of the transition, and describing the behaviour of the properties of the system in the vicinity of the transition.

The dimensionality of space and the dimensionality of the order parameter

Since our world has three spatial dimensions, it is natural, to begin with, to consider the phases and phase transitions of three-dimensional systems. However, it would be mere preCopernican prejudice to stop there. Indeed, there are many very interesting systems of lower dimensionality. Thus:

One-dimensional or quasi one-dimensional systems: threads, polymers (linear macromolecules), solids made up of chains (of atoms or molecules) with only weak coupling between chains.

Two-dimensional or quasi two-dimensional systems: films, adsorbed phases, solids made up of weakly-coupled planes (like mille-feuilles pastry).

In this last category one should include also the transitions between surface phases, which strictly speaking are two-dimensional phase transitions, even though their effects extend through a small (but finite) volume.

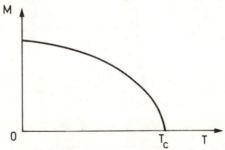

Figure 1.4. Typical variation of the order parameter M with temperature for a second-order phase transition

By contrast, in certain related problems of relativistic field theory which can be given an analogous formulation, one must work in space–time, and thus in four dimensions. By the end of this book we shall have become accustomed to considering the dimensionality d of space as a real positive variable, even though, ultimately, physical interest attaches only to certain isolated values of d, and, for critical phenomena, mainly to the value $d = 3$.

As regards the order parameter, it is not always obvious how to define it, and both experimental and theoretical difficulties can occur. Once it is defined, one of its more interesting features is its dimensionality n. In the case of the liquid-to-gas transition, the order parameter is a scalar, being a density difference; one says that its dimensionality is one. In the case of a ferromagnetic transition without anisotropic forces, the order parameter, i.e. the magnetization, is a vector having three components; one says that its dimensionality is three. Thus, the dimensionality n of the order parameter can assume various different physical values. Here again we shall find it profitable to consider n as a real positive variable; occasionally we shall even allow it zero or negative values.

Fluctuations, correlation length and critical phenomena

In the vicinity of a transition point a system has two different phases, stable, respectively, in two adjoining regions. This entails the existence of fluctuations, which as a rule govern the behaviour of the system near the transition. One example is the critical opalescence characteristic of liquid-to-gas transitions.

Let us consider for example a ferromagnetic material, with a phase diagram like that in Figure 1.2. For $T > T_c$, the stable phase is the disordered one; nevertheless, if T is close to T_c, the ordered phase is only slightly less stable, and fluctuations will lead to the appearance of locally ordered regions which persist for a certain length of time. The physics then tempts one to define for such fluctuations a correlation length ξ measuring their mean extension, and a correlation time τ measuring their mean lifetime. The length ξ and the time τ increase as one approaches the transition point, whence the critical phenomena associated with the fluctuations also become increasingly important. Roughly speaking, the correlation length ξ governs static critical phenomena, while the time τ governs the dynamic ones. For simplicity we focus attention on the correlation length ξ. When all other parameters are fixed, ξ is a function $\xi(T)$ of the temperature T, increasing as T tends to the transition point $(T \rightarrow T_c)$.

We now ask: does the correlation length diverge at T_c?

If it does diverge, i.e. if $\xi(T) \rightarrow \infty$ as $T \rightarrow T_c$, then near T_c the fluctuations are completely dominant; otherwise their influence remains partial and the phase transition occurs so to speak prematurely. This classification overlaps widely with that, due to Landau, according to the order of the transitions. If in the Landau classification the transition is second order (no discontinuity in the order parameter), then the correlation length diverges at T_c, and there is a fully developed fluctuation-dominated regime. But if the transition is first order according to Landau (finite discontinuity in the order parameter), then as a general rule the correlation length at T_c is finite, the fluctuation-dominated regime having aborted prematurely, by some greater or lesser margin. However there are some special cases where the two classifications do not strictly correspond; and for the purpose of studying critical phenomena the more directly relevant classification is that according to the behaviour of the correlation length.

Experimental results; thermodynamic and correlation measurements; static and dynamic properties

It is convenient to distinguish two different kinds of measurement: on the one hand, thermodynamic measurements, of macroscopic quantities like specific heat, magnetization, and the homogeneous susceptibility for a ferromagnet, or density and compressibility for a liquid-to-gas transition; and on the other hand, measurements of quantities related to the microscopic correlations between different points. In the magnetic case for instance, measurements of the second kind furnish information about the (two-point) correlation function for the magnetization M:

$$\Gamma(\mathbf{R}, t) = \langle M(\mathbf{R}, t)M(0, 0)\rangle$$

or about higher-order correlation functions. Observations on the scattering of X-rays, of neutrons or of light, yield information about the Fourier transform $\Gamma(q, \omega)$ of $\Gamma(\mathbf{R}, t)$. The singularities of the thermodynamic quantities and of the correlation functions have a common origin in the critical fluctuations, being linked theoretically by the fluctuation–dissipation theorem; nevertheless from the experimental point of view one is justified in distinguishing between these two classes of measurements.

Static critical properties include the thermodynamic quantities, and the correlation functions taken at equal times; dynamic critical properties include the dependence of the correlation functions on time or frequency. From the theoretical point of view one is justified in drawing this distinction by the profound differences between the results on, and the methods of approach to, these two kinds of properties. We introduce the key word 'universality' as a reminder that there exist wide classes of physically different systems showing identical static critical behaviour[†], while this is much less true of dynamic critical behaviour. This is why the present book hardly mentions the latter, in spite of the great interest and importance of some dynamic properties; static phenomena are rich enough to provide ample illustration of the methods of the renormalization group.

Critical exponents and scaling laws

Close to a critical point (i.e. to a second-order transition) one observes quantities obeying power laws with exponents that are not integers. Thus, just below T_c, the

[†] Translator's note: We shall refer to such a class of systems, with some static critical property in common, as the 'universality class' of that property. By a convenient (though linguistically reprehensible) extension we shall also refer to critical properties as having a greater or lesser degree of universality, meaning that their universality class is more or less extensive.

order parameter begins by falling proportionately to $(T_c - T)^\beta$, while just above T_c the specific heat for instance is proportional to $(T - T_c)^{-\alpha}$, the correlation length to $(T - T_c)^{-\nu}$, and the susceptibility, in the magnetic case, to $(T - T_c)^{-\gamma}$. The exponents α, β, ν, γ, and others like them, are defined in the vicinity of a given critical point, and are called critical exponents; they have been measured with experimental accuracies that leave no room to doubt that their values are indeed not integers. One is faced with singularities, because such behaviour implies that the thermodynamic quantities and the correlation functions depend non-analytically on their variables, i.e. on the temperature, the field conjugate to the order parameter, the distance, etc. In such circumstances the usual perturbation methods, which are the physicists' favourite tools, prove ineffective; singular behaviour demands a tailormade approach.

What is remarkable is that such behaviour, though singular, has some very simple characteristics. The critical exponents do not assume just any values; different systems, undergoing the most varied kinds of transitions, can be assigned to a small number of classes each specified by a certain set of values of the exponents. Moreover one observes between the critical exponents some very simple relations like $\alpha + 2\beta + \gamma = 2$, called scaling laws, whose degree of universality is even greater. Amidst the complexity a new simplicity emerges.

Homogeneity and scaling; laws of corresponding states

The scaling laws are satisfied automatically if one assumes that the singular part of the free energy $G(T - T_c, H)$, and the correlation function $\Gamma(T - T_c, \mathbf{R})$ are homogeneous functions of their arguments. (H is the field conjugate to the order parameter.) The word homogeneous is understood here in the following wide sense: we say that a function $f(x, y)$ is homogeneous if it can be written in the form

$$f(x, y) = x^\lambda \cdot g\left(\frac{y}{x^\mu}\right)$$

The hypothesis of homogeneity entails not only the scaling laws but more besides. As regards the equation of state for instance it implies that there exists a law of corresponding states, interrelating the order parameter M, its conjugate field H and the temperature $(T - T_c = t)$ through an equation of the form

$$\frac{H}{M^\delta} = h\left(\frac{t}{M^{1/\beta}}\right)$$

where β and δ are critical exponents.

Experiments have confirmed with high accuracy that such a law holds near a critical point (Stanley, 1971); and it appears that the function $h(x)$, like the critical exponents, has a very wide universality class.

We speak here of a law of 'corresponding states' because it establishes a correspondence between different states of one and the same system; this correspondence is obtained by a suitable scaling transformation. More generally, a law of corresponding states enables one to compress many results into a compact form; it is a traditionally valued way of displaying experimental or theoretical results. The approach through the renormalization groups consists precisely in the theory of such corresponding states, which manifest themselves in homogeneity properties and in laws of corresponding states.

Universality

Every scientific investigation of a real physical system must at some stage determine which are the important, dominant, or relevant parameters. It is the irrelevance of very many of the parameters which underlies the reproducibility of experiments on a given system, and the applicability of results from one system to another.

Which are the relevant parameters in a given system can depend on the class of phenomena being considered. One good example is hydrodynamics, where a set of just a few variables enables one to describe phenomena involving low frequencies and slow spatial variation. Another good example is the Landau theory of Fermi liquids.

In the theory of critical phenomena in phase transitions, the concept of universality plays an outstanding role. The range of applicability (i.e. the universality class) of each result is carefully investigated, and there emerges a whole hierarchy of degrees of universality. At the bottom of this hierarchy for instance one finds the critical temperature whose value depends on every minor detail; at the top are certain pure numbers, the critical exponents, which depend only on the most general properties of the system, like the dimensionalities of space and of the order parameter, but are independent of the nature of the transition, of the detailed features of the interaction, etc. Thus, the universality properties of critical phenomena allow one to establish correspondences between systems, and equivalence classes of wide extent; and we stress once more that with these classes there are associated sets of numbers which demonstrate in the clearest possible fashion the differences between classes.

These features (sets of numbers, extreme universality, degrees of universality) account for the present great interest in critical phenomena, which is justified not only by their practical importance, but also by the fact that they offer an exemplary field of application for certain theoretical techniques.

1.2   Introduction to the renormalization group

What lies behind the words 'renormalization group'? The answer is: the idea of a continuous family of transformations. Of the mathematical expression 'group' we retain only the connotation that the product of two transformations is defined; as a rule it would actually be more correct to refer to a semi-group, the difference bearing on the reversibility of the transformations. These transformations establish correspondences between sets of parameters defining physically different states. We refer to such a correspondence between parameters as a renormalization of parameters, a definition which follows historically established usage. In many important cases the correspondence in question does effectively amount to a change in scale, or in other words in the norm of the parameters, whence the term 'renormalization'; in such cases one observes scaling properties and laws of corresponding states, like those described in the last section in connection with the critical phenomena in phase transitions.

Accordingly, the renormalization group is a family of correspondences. Our task is to see how these can be established.

Genesis

Amongst all the variations already played on the theme of the renormalization group, it is convenient to distinguish two main versions, which for lack of a better description we shall call the first and the second version.

The first version was originated within the framework of field theory by Stueckelberg and Petermann (1953), by Gell-Mann and Low (1954) and by Bogoliubov and Shirkov (1959). They aimed at the difficulties which in field theory, and particularly in quantum electrodynamics, attend the appearance of the so-called ultraviolet divergences, stemming from large energies and momenta. In order to eliminate these divergences one is led to define bare and renormalized masses, and bare and renormalized interactions. The first version of the renormalization groups hinges on a degree of arbitrariness in the renormalization procedure; this arbitrariness stems from a cutoff parameter at high momenta, which is the reciprocal of a minimum length. When one is interested only in phenomena that should be independent of the cutoff, one is led to consider the renormalization process in the limit where the cutoff parameter tends to infinity. After a promising start this first version ran into a lean period; but now under the stimulus of the developing second version it has once again begun to flourish, in a way signalized particularly by the exploitation of the so-called Callan–Symanzik equations (Brézin, Zinn-Justin, 1974).

We must state from the outset that the present book deliberately stresses the second version, for good reasons which will be spelled out later; one sufficient reason is that the first version already possesses a large literature of books and review articles. Nevertheless it is illuminating to give a quick sketch of how the basic ideas evolved, in the context of three specific problems: quantum electrodynamics, the Kondo effect, and critical phenomena.

In quantum electrodynamics, the perturbation solution gives a series in powers of the coupling constant (the electric charge), whose individual terms diverge logarithmically (ultraviolet divergence). By re-summing the most divergent terms by the so-called parquet method, Landau and his school obtained power-law behaviour at asymptotically large momenta. Bogoliubov then showed how these results could be obtained more succinctly by using the renormalization group in its first version, the only version extant at the time.

Oddly enough the same sequence was repeated in the theory of the Kondo effect. The Kondo effect concerns the singular behaviour of magnetic impurities in metals at low energy and low temperature. Here, the coupling constant is the exchange integral between the magnetic impurity and the conduction electrons; perturbation expansions in power of the coupling constant lead to terms which diverge logarithmically. Abrikosov (1965) re-summed the most divergent terms by the parquet method; Fowler and Zawadowski (1971) reproduced the same results by aid of the first version of the renormalization group, the role of the cutoff parameter being played here by the width of the conduction band. The analogy stopped when Anderson (1970) obtained the same results by a new method inspired by the second version of the renormalization group.

The new physical idea is a systematic reduction in the number of degrees of freedom. In the Kondo effect one eliminates the electronic degrees of freedom, starting with the electronic energy levels near the band edges, and using for this purpose a renormalization of the parameters in the Hamiltonian; in this way one exploits a well-defined correspondence to construct a family of Hamiltonians each of which describes the same properties of the initial system. All these Hamiltonians contain the same cutoff parameter, which is caused to revert to its initial value after each reduction in the number of degrees of freedom; in this lies the difference from the first version, where the cutoff is eliminated from the problem by letting it tend to infinity.

Earlier than this, the same idea of a systematic reduction in the number of degrees of freedom had arisen in the theory of critical phase transitions. It is found, as to its essentials, in the 'decimation process' and the 'blocks' of Kadanoff. To be definite, Kadanoff (1966) considers a lattice of interacting spins (ferromagnetic transition); close to the critical point the correlation length far exceeds the lattice constant a which is the distance between neighbouring spins. The reciprocal of a acts as cutoff parameter for large momenta. Kadanoff proposes to take the following steps: form blocks of spins, consider the blocks to be the new basic entities, calculate the effective interactions between them, and in this way construct a family of corresponding Hamiltonians. Clearly this amounts to eliminating short-wavelength fluctuations, i.e. those with wavenumbers close to the cutoff parameter. Moreover, the formation of blocks of spins provides a very concrete model, in real space, of the systematic reduction in the number of degrees of freedom, or, more exactly in this case, of the reduction of their density in space. While it made light of all manner of difficulties affecting its practical implementation, Kadanoff's approach did 'derive' the homogeneity properties of the free energy and of the correlation function.

The construction of a reliable basis for what we call the second version of the renormalization group was reserved for Wilson, a student of Gell-Mann and a colleague of Fisher, and a field-theorist not unversed in statistical mechanics. While tackling critical phenomena, Wilson gave precise definition to and revealed the latent potentialities of the intuitive suggestions of Kadanoff (Wilson and Kogut, 1974); there has resulted a rich harvest of results which are described in this book.

Wilson was moved by his impetus to qualify the two versions of the renormalization group, perhaps somewhat rashly, as the ancient and the modern. In actual fact and as regards the theory of critical phenomena, almost all the results obtained from the second, Wilson's modern version have been rederived from the first, i.e. from Wilson's ancient version, occasionally with less effort; moreover, certain results have been obtained from the first version but not from the second. There has ensued a slight disagreement between the ancients and the moderns as to the relative merits of the two versions. This disagreement has a subjective aspect involving questions of taste, and an objective aspect involving questions of rigour and of development-potential.

The moderns, comprising mostly those who are not field theorists, consider the second version more revealing intuitively and more heuristically suggestive: they also believe it to be more general, on the strength of various recent applications to problems other than critical phenomena.

Reducing the number of degrees of freedom. Covariance under dilatation

Next, we shall elucidate the connection between the idea of a correlation length and the idea of a reduction in the number of degrees of freedom. The degrees of freedom which are effectively coupled to each other are those contained in a volume with linear dimensions of the order of the correlation length. The approximation methods which are most widely used neglect correlations between large numbers of particles, and are valid only when the correlation length is small. In practice one can deal simply and consistently with the correlations between pairs of particles; as soon as one tries to include correlations between three particles at a time, the difficulties increase considerably. This kind of approach is bound to fail when the correlation length is large.

Expressed in this language, reducing the number of degrees of freedom amounts to establishing a correspondence between one problem having a given correlation length and another where the correlation length is smaller by a certain factor, which depends on the reduction ratio. One special case which immediately springs to mind is that where the correlation length is strictly infinite to begin with and therefore remains infinite as the reduction proceeds.

Accordingly, the renormalization group establishes correspondences between systems having different correlation lengths. If, by whatever method, this sequence leads to a solvable system, then by reversing the argument one can solve the system with which one started. As regards phase transitions, the critical behaviour is found by studying small deviations from the special case of the critical point, where the correlation length remains infinite at every stage of the reduction process.

When one speaks about reducing the correlation length, the language reminds one of a symmetry operation, namely a dilatation of the unit of length, which is equivalent to a reduction in the correlation length. In other words, at the critical point a system is invariant under dilatation, while in the vicinity of the critical point dilatation symmetry is broken, to an extent which increases as one moves further away from the critical point. From this viewpoint it is the limiting case of infinite correlation length which appears simple, and it is so in a certain sense; the situation is simple when the correlation length is either very small or very large.

These remarks notwithstanding, the reduction in the number of degrees of freedom through the operations of the renormalization groups represents a very sophisticated kind of dilatation; in particular much more is involved than simple dimensional analysis. Indeed we shall see that only in a trivial (the Gaussian) case does the renormalization group reduce to dimensional analysis. One says in this trivial case that the various parameters have their normal or canonical dimensions. But in the general case one is led to define so-called anomalous dimensions, different from those suggested by naïve dimensional arguments. It is these anomalous dimensions that characterize the covariance properties of the various parameters under dilatation; they are responsible for the scaling properties, and their 'anomalous' values are responsible for the non-classical values of the critical exponents.

One should realize that there is no necessity for the dilatation to occur in real space, i.e. in the space reciprocal to momentum space. For instance, we have seen already that in the Kondo effect one works with a dilatation in energy or in time.

Fixed points in parameter space

The sequence of corresponding Hamiltonians which one obtains while reducing the number of degrees of freedom is usefully pictured as a trajectory in a space spanned by the system parameters. These are the temperature, the external fields and the coupling constants. To every point in parameter space there corresponds a Hamiltonian at a given temperature, or in other words, one definite state of a system at one definite temperature; the temperature is considered as just one parameter amongst several, all on an equal footing. All these Hamiltonians have one common cutoff parameter, since after each reduction process this parameter is caused to revert to its initial value. In order to include the entire trajectories of all the representative points, one must work with a fairly large parameter space; as a rule this means a space of high dimensionality, because the reduction process introduces new kinds of couplings. One can see already that the method will become impracticable unless it can be simplified.

For instance, consider a system undergoing a ferromagnetic phase transition. As the temperature varies, the point representing the system with its given set of parameters (external fields and coupling constants) moves along a curve in parameter space. We shall call this curve the physical line (see Figure 1.5). Further, in this parameter space we can draw surfaces S_ξ on each of which the correlation length ξ takes on a constant value. The special surface S_∞ which corresponds to an infinite correlation length is called the critical surface.

Parametrized by the temperature, the physical line of the initial system intersects these constants-ξ surfaces, because the correlation length changes with temperature; the critical temperature T_c corresponds to the point P where the line intersects the critical surface. To every point of the physical line, and issuing from that point, the method of the renormalization group assigns a trajectory of corresponding systems. The trajectory which issues from the point P is confined to the critical surface S_∞ because the correlation length remains infinite under the

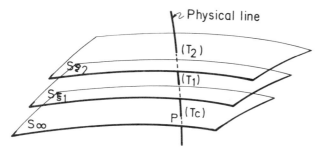

Figure 1.5. Constant–ξ surfaces in parameter space (one of these being the critical surface S_∞); also shown is the physical line of a typical system, parametrized by the temperature. On this diagram, $\xi(T_1) = \xi_1$, $\xi(T_2) = \xi_2$ (with $\xi_1 > \xi_2$ and $\xi(T_c) = \infty$

procedure reducing the number of degrees of freedom. By contrast, the trajectory starting from any other point near P will veer away from the critical surface, and will intersect surfaces S_ξ of successively lower ξ values, as the correlation length contracts under the reduction procedure.

We are now in a position to introduce the crucial concept of a fixed point. As the name implies, a fixed point is a point in parameter space which by itself constitutes its own entire trajectory. To such a point there corresponds a state of the system invariant under the operations of the renormalization group; moreover this state has a correlation length which is necessarily either infinite or zero, these being the only values that remain unchanged when reduced in any fixed proportion. Accordingly, any fixed point must lie either on S_0 or on the critical surface S_∞. We shall be interested in fixed points on the critical surface, and shall see that in certain cases the existence of such points can be shown, and their positions and catchment areas determined.

Relevant parameters and the stability of fixed points. Competition and crossover

A priori one might have expected any arbitrary behaviour of the trajectory issuing from P and inscribed on the critical surface; it might display discontinuities, zigzags, oscillations, etc. But in all cases worked out so far one observes either convergence towards a fixed point or divergence to infinity. One can then form a very intuitive picture of the situation. Let us take the most easily visualized case (see Figure 1.6) where the critical surface is two-dimensional, and picture it as a relief map, with the trajectories of the renormalization group marking the lines of steepest descent, along which water would drain. Three types of fixed point are possible: bottom-of-hollow, saddle-point (mountain pass or col), and peak. They are characterized by their stability. As a rule those maximally stable (bottom-of-hollow) correspond to ordinary critical points, those of lesser stability (saddle) to tricritical points, and so on.

14

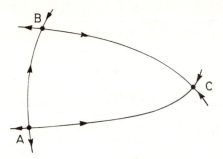

Figure 1.6. One possible pattern of
trajectories on the critical surface. A, B,
C are fixed points; A is a peak, B a saddle
point and C a hollow

In the neighbourhood of a fixed point one can construct a linearized theory by
studying the renormalization of parameters differing but little from those of the
fixed point itself. In this way one defines local axes (see Figure 1.7) by diagonalizing
a linear transformation matrix. In turn, the local axes define what are called scaling
variables, i.e. parameters which, near the fixed point, simply change in scale when
acted on by the transformations of the renormalization group. In other words
these transformations simply multiply the scaling variables by numbers s^y where s
is the dilatation of the unit of length, and y is the 'anomalous dimension' of the
scaling variable. Except for reasons of symmetry the local axes and therefore the
scaling variables usually differ for different fixed points. If, along a given local axis,
the trajectory behaves centrifugally, i.e. if its anomalous dimension is positive, then
the corresponding scaling variable is called relevant; if the behaviour is centripetal, i.e. if the anomalous dimension is negative, the variable is called irrelevant.
As a rule, the temperature is a relevant variable, in that the trajectories starting
from points on the physical line near to P, though initially they approach the
fixed point associated with P, veer away from it eventually (see Figure 1.8).

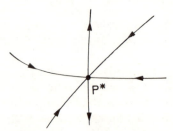

Figure 1.7. Topology of trajectories near a fixed point P*
having three local axes, two
corresponding to irrelevant fields
and one to a relevant field

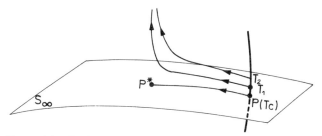

Figure 1.8. Trajectories issuing from points on a physical line which intersects the critical surface at P; P lies in the catchment area of a fixed point P*. The order of temperatures is $T_2 > T_1 > T_c$

To each fixed point there corresponds a specific critical behaviour, governed by the local scaling variables and their anomalous dimensions. Associated with each fixed point is its catchment area on the critical surface. Every system whose physical line intersects a given catchment area displays the critical behaviour appropriate to the fixed point associated with that catchment area. This yields a very picturesque visual image of the phenomenon of universality, analogous to a drainage map where the different catchment areas are separated by lines marking the watersheds. In exceptional cases, i.e. for special values of some parameter (other than the temperature), the physical line can intersect the critical surface at a point P lying on a watershed; then one finds convergence towards a fixed point having a lower degree of stability, as for tricritical points.

The competition between fixed points gives rise to the so-called crossover phenomena. It is possible for a trajectory to pass close by one fixed point before finally converging on another. This is reflected by the successive display of two different types of critical behaviour. As the temperature approaches the critical temperature, one observes first a critical regime characteristic of the first fixed point; this is followed by a transitional or 'crossover' zone, and finally, in the immediate vicinity of T_c by a second critical regime characteristic of the second fixed point, to which the trajectory finally converges.

The regions of the critical surface from which trajectories diverge to infinity are interpreted, in the case of phase transitions, as indicative of first-order transitions. In the case of the Kondo effect, such behaviour signalizes a qualitatively new kind of ground state. This list of possibilities is not exclusive.

Topological description

In the preceding subsections we have tried to give a quick sketch of the way in which the renormalization group leads one to describe the phenomena. This topological description is very alluring because in advance of any detailed calculation it enables one to visualize universality and scaling, the existence of different kinds of fixed points, crossover phenomena (i.e. passage from one critical regime to another) and so on. It even allows one to understand how critical regimes

16

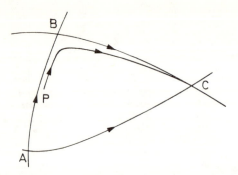

Figure 1.9. The arrangement in Figure 1.6
repeated to illustrate competition between
fixed points. The trajectory issuing from a
point P close to the ridge AB at first
approaches the fixed point B but then veers
away to C

with certain special features (logarithmic factors multiplying the power-law
expressions, apparent violations of universality, etc.) arise as consequences of
identifiable topological accidents (accidentally coincident fixed points, lines of
fixed points, etc.). To formulate this topological description we have used a verbal
analogy to relief maps showing watersheds, lines of steepest descent, saddle points,
and so on. At this purely descriptive stage there is another amusing analogy to the
stellar evolution plots in a Hertzprung–Russell diagram. There too one could
speak of catchment areas, of crossover behaviour, and of different kinds of fixed
points associated respectively with white dwarfs, neutron stars and black holes.

One great merit of the topological description of the renormalization group is
that it transforms a highly singular problem into a problem that is regular albeit in
an abstract space. We have noted the remarkable fact that in a phase transition a
small variation of a parameter suffices to trigger qualitative changes. By aid of the
trajectories in parameter space, we can now see that a small change in the initial
conditions can despatch a trajectory into very different directions, depending on
whether or not the starting point lies on the critical surface, and, if on this surface,
on whether it lies in one catchment area or in another. And all this without any
non-analiticity in the system of differential equations defining the trajectories.

Moreover, the topological description has great heuristic value. Assume that by
studying some limiting cases we have discovered the general layout of the
trajectories in certain regions of parameter space; we can then appeal to a criterion
of topological simplicity, lacking the power of proof but of great heuristic appeal,
in order to visualize various possibilities for reconciling the behaviour of different
trajectories, and to arrange these possibilities in order of increasing topological
complexity. As long as the topologically simplest case remains compatible with all
the information available about the system, it provides an acceptable working
hypothesis on which predictions can be based. If later on these predictions fail, one
can try the next simplest solution, and so on. Such a theory thus provides a

framework for approximations; its interest is not limited to cases that are soluble exactly.

Finally, note as a curiosity that the topological description links renormalization-group methods to the mathematical theory of dynamical systems. The latter aims at a unified description of many problems in hydrodynamics, chemistry, ecology, biology and in social studies; we refer to the book by René Thom (1975). The one essential difference is that the role which in dynamical systems is played by time, is, in the renormalization group, assigned to the dilatation ratio; apart from this difference, one is led in both cases to a global description of the phenomena, to an analogous classification of singularities, and to a similar understanding of universality properties. The link is worth noting, because it may amount to an indication that a new level has been reached in the development of our understanding of these matters.

References

Abrikosov, A. A. (1965). *Physics*, **2**, 5.

Anderson, P. W. (1970). *J. Phys. C*, **3**, 2436.

Bogoliubov, N. N., Shirkov, D. V. (1959). *Introduction to the theory of quantized fields*, Interscience (New York).

Brézin, E., Zinn-Justin, J. (1974). Field theoretical approach to critical behaviour, to appear in *Critical Phenomena*, vol. VI, Domb, C. and Green, M. S. (Ed.), Academic Press.

Fowler, M., Zawadowski, A. (1971). *Solid State Comm.*, **9**, 471.

Gell-Mann, M., Low, F. (1954). *Phys. Rev.*, **95**, 1300.

Kadanoff, L. P. (1966). *Physics*, **2**, 263.

Landau, L. D., Lifshitz, E. M. (1958). *Statistical Physics*, Pergamon (London).

Stanley, E. H. (1971). *Introduction to Phase Transitions and Critical Phenomena*, Clarendon Press (Oxford).

Stueckelberg, E. C. G., Petermann, A. (1953). *Helv. Phys. Acta*, **26**, 499.

Thom, R. (1975). *Structural Stability and Morphogenesis*, Benjamin (Reading, Mass.).

Wilson, K. G., Kogut, J. (1974). *Phys. Reports*, **12C**, No. 2.

CHAPTER 2
Introduction to critical phenomena in phase transitions

> 'Human understanding progresses naturally from individuals to
> species, from species to families, from closely related families to
> those more remote, and at each level it constructs a science'
>
> d'Alembert

In the nineteenth century (van der Waals, 1873) all manner of more or less special models were exploited in order to account for the various phase transitions that had been observed, starting with the most common (liquid-to-gas, magnetic, order–disorder in alloys, etc.). Gradual recognition of the universality properties justified the choice of a model which is convenient yet representative. In Section 2.1 we first recall some models that have played especially important roles (like spin-lattice and vertex models) and then proceed to the Ginzburg–Landau theory, sometimes called the 'n-vector model', which expresses the partition function as a functional integral over all possible spatial variations of the order parameter. This will be followed by the definition of the main critical exponents.

Section 2.2 describes the Gaussian model; though this is trivially soluble and has pathological features, it nevertheless constitutes a very important special case. The Landau theory of critical phenomena, leading to the so-called classical values of the critical exponents, is given in Section 2.3. In Section 2.4, the Ginzburg criterion allows us to determine the conditions under which the Landau theory applies, and highlights the crucial role of the space dimensionality d. This leads to the definition of a characteristic dimensionality d_c, such that for $d > d_c$ the critical exponents assume their classical values, while for $d < d_c$ there exists a critical region, with a width determined by the Ginzburg criterion, where the critical exponents assume non-classical values. Section 2.5 contains various generalizations of the Landau theory and the Ginzburg criterion to long-range forces and to critical points of special kinds, like tricritical points.

Finally in Section 2.6 we study the part played by the dimensionality n of the order parameter; we present the special case of $n = 0$ of random walks, (the excluded-volume problem for polymers), describe the limiting cases $n = -2$ and $n = \infty$, and sketch a systematic classification of critical phase-transition phenomena according to the values of the dimensionalities d and n.

2.1 Ginzburg–Landau theory

Before describing the so-called Ginzburg–Landau theory it will be useful to review briefly some important models.

Heisenberg and Ising Models

The Heisenberg model consists of quantized spins located at the sites of a lattice and coupled in pairs by exchange interactions; the Hamiltonian is

$$\mathcal{H} = \sum_{ij} J_{ij}\mathbf{S}_i \cdot \mathbf{S}_j$$

and may be rewritten as

$$\mathcal{H} = \sum_{ij\alpha} J_{ij}^{\alpha} S_i^{\alpha} S_j^{\alpha}$$

where the indices i, j specify lattice sites, and the index α the components, x, y and z, of the spins. By setting $J_{ij}^z = 0, J_{ij}^x = J_{ij}^y \neq 0$, one obtains the so-called XY model; by making $J_{ij}^x = J_{ij}^y = 0, J_{ij}^z \neq 0$, one obtains the Ising model.

The Heisenberg and Ising models have been very important in the history of critical phenomena. The exact solution of two-dimensional Ising models has played a major part, first by demonstrating the existence of power laws with non-integral critical exponents different from the classical values, and secondly by proving the universality of these exponents with respect to various characteristics of the system, like its lattice structure.

Many numerical calculations on the Heisenberg and Ising models in three dimensions have confirmed the universality of their static critical properties with respect to the lattice structure, the range of forces (if finite) and the magnitude of the spins. By contrast, Heisenberg models differ from Ising models in their critical exponents, which shows the importance of the number of components of the order parameter: $n = 3$ for Heisenberg but $n = 1$ for Ising models. Moreover the exponents of the Ising model are different in three dimensions than in two, which shows the importance, too, of the dimensionality of space.

Other models

Let us note the generalizations of the Ising model due to Ashkin, Teller and Potts. Abandoning the idea of spins as such, one assigns to each lattice site m possible internal states, and postulates, for instance, that there is a non-zero interaction between two neighbouring sites if they are in different internal states, but no interaction if they are in the same state. In this way one can arrive at new kinds of order, i.e. alignment, which are useful for instance in the study of nematic liquid crystals.

Under the name vertex models one includes lattice models where the couplings between sites have directional properties indicated by arrows; the energy associated with each site depends on the configuration of arrows on the adjacent bonds. Depending on the energies assigned to different configurations, one meets, as special cases, certain models for ice, for ferroelectrics and for antiferroelectrics. Some of these models have been solved for dimensionality $d = 2$, in particular the famous six-vertex or Baxter model.

Originally, the philosophy suggested by these exact solutions seemed to be the opposite of that suggested by the Heisenberg and Ising models. While the latter displayed wide-varying universality properties, the former behaved very differently from each other; the Baxter model even possesses an exponent which varies continuously as a function of the configuration energies. This apparent paradox is now fairly well understood; anticipating somewhat, one says that the Heisenberg and Ising models are simple systems, while the vertex models are complex; complex systems are simple systems that have been perturbed, and the effect of some perturbations is to restrict the universality class ('marginal' perturbation in the case of the Baxter model; see Chapter 13).

Ginzburg–Landau theory

As regards static critical properties, it has been found that the structure of the lattice, the range of the force (if finite) and the magnitude of the spins are all irrelevant; hence it would seem convenient to choose a model with classical spins (which assume continuously variable values), and to define these as no longer confined to discrete lattice sites, but distributed continuously through space. More precisely, one defines, at every point $\mathbf{x}$ of a d-dimensional space, a field variable $\mathbf{M}(\mathbf{x})$ having n components.

The partition function Z, which is central to every calculation of thermodynamic properties, can then be written as a functional integral over all spatial variations of the field variable:

$$Z = \int \mathscr{D}\mathbf{M}(\mathbf{x}) \exp\left[-\beta \int F_{\mathrm{L}}\{\mathbf{M}(\mathbf{x})\}\, \mathrm{d}\mathbf{x} \right] \tag{2.1}$$

$F_{\mathrm{L}}\{\mathbf{M}(\mathbf{x})\}$ is the local free-energy density, and $\beta = 1/T$ the reciprocal temperature. The choice of F_{L} then defines the problem. The standard choice is

$$F_{\mathrm{L}}\{\mathbf{M}(\mathbf{x})\} - F_0 = A \sum_{i=1}^{n} M_i^2(\mathbf{x}) + B\left(\sum_{i=1}^{n} M_i^2(\mathbf{x}) \right)^2 + K \sum_{i=1}^{n} \sum_{\alpha=1}^{d} \left(\frac{\partial M_i}{\partial x_\alpha} \right)^2 \tag{2.2}$$

being the start of a series expansion in powers of $\mathbf{M}(\mathbf{x})$ and of its derivatives, one which contains no odd terms (there is symmetry under the parity operation $M \rightleftharpoons -M$), and where the form of the gradient term implies, as will become clear later, that the forces are short-range.

There is a simple way to see how, starting from the general definition of Z:

$$Z = \mathrm{Tr} \exp(-\beta \mathscr{H})$$

one can arrive at the formula (2.1). Let us evaluate the trace in two steps, summing first over all complexions (microscopic configurations) that are compatible with a prescribed spatial variation of the field variable $\mathbf{M}(\mathbf{x})$. Let $W\{\mathbf{M}(\mathbf{x})\}$ be the number of such complexions; then Z becomes

$$Z = \int \mathscr{D}\mathbf{M}(\mathbf{x}) W\{\mathbf{M}(\mathbf{x})\} \exp[-\beta E\{\mathbf{M}(\mathbf{x})\}]$$

$$= \int \mathscr{D}\mathbf{M}(\mathbf{x}) \exp[-\beta F\{\mathbf{M}(\mathbf{x})\}]$$

where the free energy $F\{\mathbf{M}(\mathbf{x})\} = E\{\mathbf{M}(\mathbf{x})\} - T \log W\{\mathbf{M}(\mathbf{x})\}$, the sum of an energy and an entropy term, may be written as

$$F\{\mathbf{M}(\mathbf{x})\} = \int d\mathbf{x} F_L\{\mathbf{M}(\mathbf{x})\}$$

being a sum over all space of a local free-energy density.

More generally, in presence of an external field coupled to the field variable (e.g. in the case of a ferromagnet, the magnetic field coupled to the magnetization), one has

$$Z(T, H) = \int \mathscr{D}\mathbf{M}(\mathbf{x}) \exp\left[-\beta\left(\int d\mathbf{x} F_L\{\mathbf{M}(\mathbf{x})\} - \mathbf{H} \int d\mathbf{x} \mathbf{M}(\mathbf{x})\right)\right]$$

Then the Gibbs free energy (thermodynamic potential) $G(T, H)$, a function of temperature T and field H, is given by

$$G(T, H) = -\frac{1}{\beta} \log Z(T, H)$$

This definition makes G an extensive variable proportional to volume; usually we shall be interested in the corresponding density, i.e. in G defined for unit volume.

In the expression (2.2) for the local free energy density, the coefficients A, B, K embody contributions from both energy and entropy; the energy favours order and the entropy favours disorder. In particular, the contribution of the energy to the coefficient A is usually negative, and that of the entropy positive; thus A will turn out positive at high and negative at low temperatures, and it is this variation of A with temperature that triggers the phase transition.

Definition of the main critical exponents

Let T_c be the phase-transition temperature. In order to describe the behaviour near T_c of certain thermodynamic variables and of the correlation function, one defines six main critical exponents: $\alpha, \beta, \gamma, \delta, \eta, \nu$.

The exponent β refers to the variation of the order parameter in the low-temperature phase (e.g. for ferromagnets, the spontaneous magnetization)

$$M \sim |\Delta T|^\beta, \qquad \Delta T = T - T_c$$

(Take care not to confuse the exponent β with the reciprocal temperature.)

The specific heat C and the susceptibility χ involve the exponents α and γ:

$$C \sim \frac{1}{|\Delta T|^\alpha}, \qquad \chi \sim \frac{1}{|\Delta T|^\gamma}$$

At $T = T_c$ the relation between field and magnetization leads one to define the exponent δ by

$$H \sim M^\delta$$

and the long-distance behaviour of the correlation function $\Gamma(R)$ defines the exponent η by

$$\Gamma(R) \sim \frac{1}{R^{d-2+\eta}}$$

Finally the exponent ν refers to the correlation length ξ:

$$\xi \sim \frac{1}{|\Delta T|^\nu}$$

For a long time one made a distinction as regards the exponents α, γ, ν, between their high-temperature values $(T > T_c)$ and their low-temperature values, $(T < T_c)$, identified by primes; but it seems now that the primed and unprimed values are always equal, so that the distinction has been abandoned. However, it would be wrong to think that there is perfect symmetry between behaviour above and below T_c. On the one hand, the coefficients in front of the powers differ in the two cases, which could cause experimental results to be misinterpreted as implying that the exponents themselves differ. On the other hand, we shall see later, in Section 6.3, that if a gauge-variable† is present (as for a broken continuous symmetry, $\eta > 1$), then in zero field the susceptibility and the correlation length are strictly infinite throughout the low-temperature phase $(T < T_c)$.

Between the six critical exponents defined above there hold four relations called scaling laws. The first three involve only the exponents themselves:

Rushbrooke's scaling law: $\alpha + 2\beta + \gamma = 2$
Widom's scaling law: $\gamma = \beta(\delta - 1)$
Fisher's scaling law: $\gamma = (2 - \eta)\nu$

while the fourth involves also the dimensionality of space:

Josephson's scaling law: $\nu d = 2 - \alpha$

These laws were originally derived as inequalities, and have a greater degree of universality than the exponents themselves. By virtue of these four relations (to the extent that they can be proved or are accepted), the problem of determining the six critical exponents reduces to determining any two of them.

Beyond the calculation of these exponents, the theory of static critical phenomena is interested in determining asymptotic forms of the equation of state,

$$\frac{H}{M^\delta} = f\left(\frac{\Delta T}{M^{1/\beta}}\right) \tag{2.3}$$

---

† Translator's note. When the order parameter is complex, we refer to its phase (as opposed to its modulus) as a 'gauge'-variable. An obvious alternative is 'phase' variable, which we avoid in order to avoid confusion with the word 'phase' in the usual thermodynamic sense.

and of the correlation function

$$\Gamma(R, T) = \frac{1}{R^{d-2+\eta}} \cdot g\left(\frac{R}{\xi}\right) \tag{2.4}$$

and also in correlation functions of higher order, or in the presence of external fields, etc.

Finally we note that the exponents have been defined, above, in a way that is especially well adapted to displaying experimental results; but from a theoretical point of view it will prove more convenient later on to define other exponents, namely the anomalous dimensions of the fields, in terms of which the scaling laws become simpler and more transparent.

Back to universality

Starting from the Ginzburg–Landau formulation we shall derive the universality of the static critical properties; more precisely, we shall see that many parameters in the problem fail to affect these properties, i.e. are irrelevant. By contrast, the degree of universality of dynamic critical phenomena is far more restricted, which explains why they are far less well understood and why this book does not deal with them. Returning to static properties, one might ask to what extent the results on the Ginzburg–Landau model are applicable to other models like the Ising and Heisenberg models, and so on.

Though it has not yet been rigorously proved, the great majority of experts agree that whether the space is discrete or continuous is irrelevant to the critical properties; this appeal to authority aims to sidestep an enumeration of the arguments, which will appear later. As regards the distinction between field variables which assume discrete or continuous values (finite or infinite spins), and more generally between quantum or classical variables, this too is irrelevant provided the transition occurs at a finite temperature $T_c \neq 0$. But if the transition occurs at zero temperature, which for low dimensionality could be enforced by the fluctuations, then the critical behaviour could be influenzed by the quantized nature of the variables; what happens in such cases is that the zero-point energy of the lowest normal modes is no longer negligible compared to the temperature, and the degree of universality of the static properties then comes to depend on the more restricted universality of the dynamical properties.

2.2 The Gaussian model

The Gaussian model is defined by the following choice of the local free-energy density:

$$F_L\{M(\mathbf{x})\} - F_0 = A \cdot (M(\mathbf{x}))^2 + K(\nabla M)^2 \tag{2.5}$$

Here we have retained only one component of $M(\mathbf{x})$; if F_L is quadratic in $\mathbf{M}$ and its derivatives, each component can be integrated independently of the others, and the number of components then appears simply as a factor multiplying the thermodynamic potential.

The Gaussian model is exactly soluble, since the formula for F_L involves only Gaussian integrals. It is convenient to work in the space of the Fourier transform $M(\mathbf{k})$ of $M(\mathbf{x})$, since the different Fourier components decouple. Then one has

$$F\{M(\mathbf{k})\} - F_0 = \sum_{\mathbf{k}} F_{\mathbf{k}} \quad \text{where} \quad F_{\mathbf{k}} = (A + Kk^2)|M(\mathbf{k})|^2$$

For convenience one normalizes the energy, setting $K = \dfrac{1}{2}$, $A = \dfrac{r_0}{2}$, and rewrites this as

$$F_{\mathbf{k}} = \tfrac{1}{2}(r_0 + k^2)|M(\mathbf{k})|^2 \tag{2.6}$$

The partition function becomes

$$Z = \prod_{\mathbf{k}} \int dM(\mathbf{k}) \exp(-\beta F_{\mathbf{k}})$$

which is a Gaussian integral leading to

$$G \sim T \int d\mathbf{k} \log(r_0 + k^2)$$

This expression is well-defined only if r_0 is positive (high temperatures); since r_0 varies smoothly with temperature, one chooses to write, for simplicity,

$$r_0 = T - T_0$$

where T_0 is the transition temperature. The low-temperature phase $T < T_0$ is not well-defined because the integral diverges; this is a pathological feature of the Gaussian model. Near T_0, one can write G as

$$G - G_0 \sim (\Delta T)^{d/2}(1 + \Delta T + \cdots) + \Delta T(1 + \Delta T + \cdots) \tag{2.7}$$

where $\Delta T = T - T_0$. The specific heat becomes

$$C \sim T\frac{\partial^2 G}{\partial T^2} \sim (\Delta T)^{(d/2)-2}(1 + (\Delta T) + \cdots) + (\Delta T)^0(1 + \Delta T + \cdots) \tag{2.8}$$

For $d < 4$, the first (singular) term dominates on the right and one finds

$$\alpha = 2 - \frac{d}{2} = \frac{4-d}{2} \tag{2.9}$$

For $d > 4$ the specific heat no longer diverges at $T = T_0$, and the regular term becomes dominant. For $d = 4$ the specific heat diverges logarithmically:

$$C \sim \log(\Delta T)$$

To calculate the correlation function $\Gamma(k)$, we note that

$$\Gamma(k) = \langle |M(\mathbf{k})|^2 \rangle$$

and that $\langle |M(k)|^2 \rangle$ can be obtained from (2.6) by appeal to the equipartition of energy, which holds for Gaussian integrals:

$$\Gamma(k) = \frac{T}{r_0 + k^2} \tag{2.10}$$

From this, the values of the exponents η, γ and v are

$$\eta = 0, \gamma = -1, v = \tfrac{1}{2},$$

$\Gamma(k)$ assuming the homogeneous form

$$\Gamma(k) = \frac{1}{k^2} g(k\xi) \tag{2.11}$$

Fisher's scaling law $\gamma = (2 - \eta)v$ is then obeyed; indeed it is implicit once $\Gamma(k)$ assumes a homogeneous form like (2.11). Josephson's scaling law $vd = 2 - \alpha$ also holds, at least for $d < 4$; for $d > 4$ it runs into the difficulty that the value of α is somewhat arbitrary, since in the Gaussian model the low-temperature phase is not well defined. It is clear in any case that the dimensionality $d = 4$ constitutes a divide between two different regimes, on account of the infrared behaviour of the integral

$$I = \int_0^{} \frac{k^{d-1} \, dk}{[(\Delta T) + k^2]^2}$$

which enters into the calculation of the specific heat. When $d > 4$ the integral converges for $\Delta T = 0$; when $d = 4$ it diverges logarithmically; when $d < 4$ the infrared divergence at small values of k allows the dominant part to be written as

$$I \sim \int_{(\Delta T)^{1/2}}^{} \frac{k^{d-1} \, dk}{k^4} \sim (\Delta T)^{(d-4)/2}$$

In the following we shall often need to refer back to these features of the Gaussian model.

2.3 Landau theory

The Landau theory of critical phenomena is based on an approximation; in calculating any thermodynamic quantity, the approximation consists in replacing the sum over all possible variations of the field variable $M(x)$ by the largest single contribution, namely that which maximizes the integrand. The underlying principle is therefore closely similar to that of the geometrical-optics approximation to wave propagation, and to the semi-classical approximation in quantum mechanics. What is surprising here is that the Landau theory yields values for the critical exponents that are exact in spaces of higher than a certain characteristic dimensionality.

Let us consider the standard choice of local free-energy density,

$$F_2 - F_0 = A \cdot M^2 + B \cdot M^4 + K(\nabla M)^2 \tag{2.12}$$

where $A = A'(T - T_0)$, $B > 0$ and $K > 0$ (a wider family will be considered in Section 2.5).

In calculating the partition function

$$Z = \int \mathscr{D} M(\mathbf{x}) \exp\left(-\beta \int F_L \, d\mathbf{x}\right)$$

the principle of the approximation is to find the function $M(\mathbf{x})$ which maximizes the integrand, i.e. minimizes $\int F_L \, d\mathbf{x}$. The gradient term is positive definite and imposes the condition that $M(\mathbf{x}) = M$ is constant in space; the value of M is then obtained by minimizing $AM^2 + BM^4$ (see Figure 2.1).

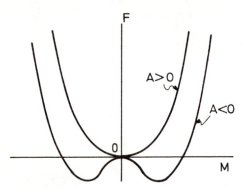

Figure 2.1. Plots of the function $F = F_L - F_0$ $= AM^2 + BM^4$ for $A > 0$ and $A < 0$, with B positive. In Landau theory, the position of the minimum of F determines the mean magnetization (which tends to zero continuously as A tends to zero from below)

This entails

For $T > T_0$ $\qquad M = 0$

for $T = T_0$ $\qquad M^2 = \dfrac{A'}{2B}|T - T_0|$

whence the value $\beta = \frac{1}{2}$ for the exponent β.

For the thermodynamic potential this yields

$$G - G_0 = -\frac{A'^2}{4B}(T - T_0)^2 \qquad \text{for} \quad T < T_0$$

$$= 0 \qquad \text{for} \quad T > T_0$$

whence the value $\alpha = 0$.

In the presence of an external field coupled to $M(\mathbf{x})$ the same variational method gives the values $\gamma = 1$ and $\delta = 3$, and the equation of state as

$$\frac{H}{M^{\delta}} = 4B + 2A'\left(\frac{\Delta T}{M^2}\right) \tag{2.13}$$

As regards the correlation function, the approximation of Landau theory consists in neglecting the quartic term on the right of (2.12); under these conditions one obtains for the correlation function the same expression (2.10) and (2.11) as in the Gaussian model in the last section. For the exponents referring to the correlation function, one finds in this way the Gaussian values $\eta = 0$ and $\nu = \frac{1}{2}$.

These values $\beta = \frac{1}{2}, \alpha = 0, \gamma = 1, \delta = 3, \eta = 0$, and $\nu = \frac{1}{2}$, are called the classical values for the critical exponents. They were obtained repeatedly in the past within the framework of the theories that could be constructed at the time, all of the mean-field type like the van der Waals theory of the liquid-to-gas transition, the Weiss theory of ferromagnetism, and the Bragg–Williams theory for order–disorder transitions in alloys. The Landau theory is the most succinct formulation of theories of this kind.

Note that the classical values have an impressive degree of universality, being independent even of the dimensionalities of space and of the order parameter. This entails that though the classical values obey the three scaling laws of Rushbrooke, Widom and Fisher, they necessarily violate Josephson's law which involves the space dimensionality d:

$$\nu d = 2 - \alpha$$

except, that is, for one particular value of d which, in view of $\nu = \frac{1}{2}$ and $\alpha = 0$, happens to be $d = 4$.

Josephson's scaling law links the exponent ν, referring to the correlation function, and the exponent α, referring to the thermodynamic potential. But the methods by which we have so far calculated these two quantities are quite patently schizophrenic; in calculating the correlation function we neglected the quartic term in F_{L}, while in calculating the thermodynamic potential we practically ignored the gradient terms. Had we neglected the quartic term systematically, then we would have been left with the Gaussian model having, (when $d < 4$), $\alpha = \dfrac{4 - d}{2}$ instead of $\alpha = 0$, and in that case Josephson's law would have been satisfied.

We shall elucidate the limits of validity of the Landau theory in the next section, but some limitations can be seen already by simple comparison with the Gaussian model: in some sense the specific heat contains a thermodynamic contribution, plus another contribution due to the Gaussian fluctuations. If the latter is negligible compared to the former, (as happens when $d > 4$, since in that case α (Gaussian) $< \alpha$ (classical)), then the internal inconsistency of the Landau theory is unimportant and the theory is exact, in the sense that it yields the dominant terms and the correct critical exponents. By contrast, when $d < 4$ the contradictive matters and a more sophisticated theory is needed.

Accordingly, as d decreases through the value $d = 4$, one moves from a regime ($d > 4$) where Josephson's law fails, into another regime ($d < 4$) where, as we shall see, it holds; the characteristic value $d = 4$ is that dimensionality for which Josephson's law is obeyed by exponents having their classical values. The Ginzburg criterion, derived in the next section, shows that this passage from one regime to the other is actually continuous in the following sense: for $d < 4$, the non-classical values of the exponents are observed only inside a critical region whose width shrinks to zero as the dimensionality tends to 4.

2.4 The Ginzburg criterion

The Ginzburg criterion estimates the width of the critical region by assessing the importance of fluctuations around the mean-field values. Outside the critical region fluctuations are negligible and mean-field theory applies; inside the critical region, i.e. near T_c, mean-field theory fails qualitatively. We shall show that for the standard choice (2.12) of free energy, the Ginzburg criterion takes the form

$$B \sim (\Delta T)^{(4-d)/2} \tag{2.14}$$

where B is the coefficient of the quartic term and ΔT the width in temperature of the critical region.

The Ginzburg criterion is a special case of a very general kind of relation between the value of a coupling constant and the temperature interval over which that type of coupling has an appreciable effect on the critical properties. In contrast to the critical exponents which have a very high degree of universality and are, in particular, independent of the coupling constants, the temperature intervals where a given type of critical behaviour is observable depend on these coupling constants very strongly indeed, as does the value of the transition temperature itself. In the Ginzburg criterion (2.14) the crossover exponent $\phi = \dfrac{4-d}{2}$ depends only on the dimensionality of space; in fact the criterion embodies a very simple dimensional relationship which underlines all the many proofs that have been given.

Derivation of the Ginzburg criterion

We start with the underlying dimensional argument. From (2.12) we construct the expression $\beta_c(F_L - F_0)$, where $\beta_c = 1/T_c$, and re-scale M with a view to simplifying the coefficient of the gradient term; the scale of M can be chosen arbitrarily, since a change in it leads only to an extra additive and non-singular contribution to the thermodynamic potential. Thus we arrive at

$$\beta_c(F_L - F_0) = \frac{r_0}{2}M^2 + u_0 M^4 + \frac{1}{2}(\nabla M)^2 \tag{2.15}$$

where $r_0 \sim T - T_0$ and $u_0 \sim B$. A dimensional analysis of the coupling constants yields

$$[r_0] = [L]^{-2}, \qquad [u_0] = [L]^{d-4}$$

Hence by prescribing r_0 and u_0 one introduces into the problem two characteristic lengths;

$$\xi_G = (r_0)^{-1/2}, \qquad \xi_1 = (u_0)^{1/(d-4)}$$

ξ_G, which is the Gaussian correlation length, rises as the temperature drops, and it becomes equal to ξ_1, which is practically temperature-independent, when

$$u_0 \sim r_0^{(4-d)/2}$$

or in other words when

$$B \sim (\Delta T)^{(4-d)/2}$$

Near T_c there appears a new characteristic length, the critical correlation length ξ, involving both ξ_G and ξ_1 in proportions which will determine the exponent ν.

Other derivations are possible, perhaps more illuminating physically but basically identical. Thus, in the ordered phase $T < T_c$ one can evaluate the width of the probability distribution $P(M)$ for the magnetization in a volume whose radius equals the Gaussian correlation length, and compare this width to the average value of the magnetization as given by Landau theory; this amounts to comparing $\Gamma(R = \xi)$ with $\langle M \rangle^2$, or $\left(\dfrac{1}{\xi_G}\right)^{d-2}$ with $\dfrac{\Delta T}{B}$, or in other words to constructing the ratio $\dfrac{B}{(\Delta T)^{(4-d)/2}}$. All these quantities are evaluated within Landau theory, and in particular ΔT is referred to as the mean-field transition temperature; when the ratio becomes of order unity, the approximations of the Landau theory become qualitatively unreliable.

Perhaps the most physical argument is the one alluded to at the end of Section 2.3; it concerns the specific heat. The specific heat as calculated in Landau theory has a discontinuity at T_c, due to the appearance of the spontaneous magnetization below T_c. To see this, note that $C = \dfrac{dE}{dT}$, and that the energy E contains a contribution proportional to $\langle M \rangle^2$; but in Landau theory $\langle M^2 \rangle \sim \dfrac{T_c - T}{B}$ when $T < T_c$, and for the specific heat this leads to a discontinuity ΔC at T_c,

$$\Delta C \sim \frac{1}{B} \tag{2.16}$$

Further, $\langle M^2 \rangle$ is obtainable from the correlation function:

$$E \sim \langle M^2 \rangle = \Gamma(R \to 0) \tag{2.17}$$

whence, using the Gaussian correlation function, one finds

$$E \sim \left(\frac{1}{\xi}\right)^{d-2} \sim (\Delta T)^{(d-2)/2}$$

This in turn implies a Gaussian contribution to the specific heat proportional to $(\Delta T)^{(d-4)/2}$, which must be compared to ΔC. The comparison then yields the Ginzburg criterion.

This last proof, involving the Gaussian value $\alpha = \dfrac{4-d}{2}$, has the advantage of displaying clearly its equivalence to the preceding argument. However, we note in passing that while the functions $\Gamma(R \to 0)$ and $\Gamma(R = \xi)$ are proportional to each other in the Gaussian case, they are not so in general; by definition one has

$$\Gamma(R \to 0) \sim (\Delta T)^{1-\alpha} \tag{2.18}$$

$$\Gamma(R = \xi) \sim \left(\frac{1}{\xi}\right)^{d-2+\eta} \sim (\Delta T)^{(d-2+\eta)\nu} \tag{2.19}$$

and, in view of the scaling laws for $d < 4$, the exponents will be equal only if $\gamma = 1$, which is the Gaussian value but is not generally valid.

Width of the critical region

The Ginzburg criterion yields, in a very suggestive way, the characteristic dimensionality d_c below which Landau theory fails because of its internal contradictions. It shows, further, that fluctuations increase in importance as the dimensionality drops, which is reflected in an increasing width of the critical region for $d < d_c$.

But how far does the Ginzburg criterion lend itself to a quantitative calculation of the width of the critical region? This question calls for a series of comments.

First, note that in the form (2.14) of the Ginzburg criterion we have significantly omitted to display the proportionality constant. Though expressions for it can be found in the literature, one should note that the width of the critical region is not a rigorously defined quantity; there is a gradual transition from one regime to another. Moreover the width of the critical region has very little universality; in particular, the Ginzburg criterion was derived above within the Ginzburg–Landau theory, whose temperature dependence arises wholly from the linear variation of the coefficient A. In a real system, if the critical region is wide, then an expansion about T_0 up to linear terms only is unwarranted, and one cannot observe a crossover from a critical to a Landau-theory regime like that described in Section 2.3. Such a crossover is clearly observable only if the critical region is narrow, $\dfrac{\Delta T}{T_c} \ll 1$. Further, one runs into the problem that the transition temperature T_c shifts away from its mean-field value; but the Ginzburg criterion assesses the width relative to T_c, because the point where mean-field theory fails is estimated in terms of the mean-field quantities themselves. This matters little provided the shift of T_c is much smaller than the width of the critical region, i.e. provided $|T_0 - T_c| \ll \Delta T$; but it becomes a source of serious trouble as the dimensionality drops.

Having made these observations about quantitative calculations of the width of the critical region, let us use some examples to illustrate the functional forms arising from the Ginzburg criterion.

One very neat test of the exponent $\dfrac{4-d}{2}$ comes from Domb's (1973) numerical calculations on the excluded-volume problems for polymers. We shall see later how this problem can be formulated so as to correspond to an order parameter with number of components zero, $n = 0$. We restate the problem briefly: let a chain having N links extend along a lattice, and let R_N be the distance between its ends; if there is no interaction between the links, one has a random walk problem and finds for large N

$$\langle R_N^2 \rangle \sim N \tag{2.20}$$

If there is a repulsive interaction w between two links placed on the same site, Domb finds that to a very good approximation the ratio $\dfrac{\langle R_N^2 \rangle}{N}$ is a function only of the product $wN^{1/2}$:

$$\frac{\langle R_N^2 \rangle}{N} = f(wN^{1/2}) \tag{2.21}$$

As $x \to 0, f(x) \to$ constant, (Gaussian limit); as $x \to \infty, f(x) \sim x^{2(2v-1)}$, (critical limit); the two regimes meet where $wN^{1/2} \sim 1$. We shall see later that the analogy between this problem and that of phase transitions in Ginzburg–Landau theory is

$$w \leftrightarrow B, \qquad N \leftrightarrow (\Delta T)^{-1}, \qquad \langle R_N^2 \rangle \leftrightarrow \xi^2$$

Hence the criterion involving $wN^{1/2}$ identical to the Ginzburg criterion (2.14), the exponent $\frac{1}{2}$ being just $\dfrac{4-d}{2}$ with $d = 3$; (the calculation applies to three-dimensional space). Note that, as predicted, this exponent is not affected by the value of n, (while the exponent v is).

The Ginzburg criterion allows a very useful discussion of how the range of forces affects the width of the critical region. Here one is interested in forces of long but finite range, proportional for instance to $\exp(-R/R_0)$, with R_0 specifying the range; forces of infinite range, i.e. those falling like powers of R, will be discussed elsewhere. Other things being equal, a change in the range R_0 will be reflected by a change in the coefficient K which enters the local free-energy density (2.12):

$$K \sim R_0^2 \tag{2.22}$$

One can see this after realizing that the contribution to the energy from a variation of the order parameter in space must be measured by the ratio of the typical range of this variation to the range of forces. On rescaling M as we did to obtain (2.15), the coefficients r_0 and u_0 vary with R_0 like

$$r_0 \sim \frac{1}{R_0^2}, \qquad u_0 \sim \frac{1}{R_0^4}$$

whence, finally, the width of the critical region varies with R_0 like

$$\Delta T \sim \left(\frac{1}{R_0}\right)^{2d/(4-d)} \qquad (2.23)$$

The longer the range R_0, the narrower the critical region; this is intuitive, since fluctuations are reduced as the interaction range increases.

In liquid Helium-4, the range of forces is comparable to the interatomic distance, and the critical region around the superfluid transition is very wide; this makes Helium-4 an ideal system for measuring critical properties. By contrast, though superconductors are systems very similar to Helium-4, nevertheless their critical region is extremely narrow, since the range of forces is much larger (comparable to the radii of Cooper pairs); so much so that any observation of their true critical properties necessitates a *tour de force*. Actually, for superconductors the expression (2.23) is ill adapted to the problem because the coefficients A, B, K in (2.12) are not independent; the useful expression is

$$\frac{\Delta T}{T_c} \sim \left(\frac{T_c}{E_F}\right)^{2(d-1)/(4-d)} \qquad (2.24)$$

where E_F is the Fermi energy of the conduction electrons. Note how sensitive to d is the width of the critical region when T_c/E_F (which is always small) is held constant.

Whether the range is finite or infinite, the rule is that increasing it tends to reduce fluctuations and, consequently, the width of the critical region. Accordingly, these widths are either zero or very small in ferroelectrics, in certain structural phase transitions dependent on long-range elastic forces, in superconductors, and so on. Where the width though small is still measurable, one observes unambiguously the crossover between a mean-field and a critical region.

2.5 Landau theory and generalized Ginzburg criterion

We shall consider two kinds of generalizations. First, to local free-energy densities of the type

$$F_L - F_0 = AM^2 + B_p M^p + \text{gradient terms} \qquad (2.25)$$

the standard choice being the special case $p = 4$ which was studied in Sections 2.3 and 2.4; this will allow us to discuss tricritical points, etc., in addition to ordinary critical points. Second, we shall consider the case when the forces are of infinite range but constant sign, i.e. of the type

$$J(r) \sim \frac{1}{r^{d+\sigma}} \qquad (2.26)$$

with positive σ; (for $\sigma < 0$ the thermodynamic limit does not exist).

The contribution to the free energy from a simple harmonic variation of the field variable,

$$M(\mathbf{r}) \sim M_k \cos(\mathbf{k}.\mathbf{r})$$

is given by

$$F_k \sim M_k^2 \int J(r)[1 - \cos(\mathbf{k}.\mathbf{r})] \, d\mathbf{r}$$

For $\sigma > 2$, if the cosine is expanded in powers of k, the resultant integral converges, and for small k one obtains $F_k \sim k^2 M_k^2$; but for $0 < \sigma < 2$, the integral diverges as $r \to \infty$, (ultraviolet divergence), and for small k one finds $F_k \sim k^\sigma M_k^2$. We shall now derive rather quickly all the results of the preceding sections, but for arbitrary values of σ, $(0 < \sigma < 2)$; short-range forces are covered by the special case $\sigma = 2$.

The Gaussian model is trivially soluble and has the exponents

$$v = \frac{1}{\sigma}, \qquad \eta = 2 - \sigma, \qquad \gamma = 1$$

and $\alpha = 2 - d/\sigma$, which is well-defined for $d < 2\sigma$ (see Section 2.2). The generalized Landau theory yields

$$v = \frac{1}{\sigma}, \qquad \eta = 2 - \sigma, \qquad \gamma = 1$$

exactly like the corresponding Gaussian model; in addition it yields

$$\alpha = \frac{p-4}{p-2}, \qquad \beta = \frac{1}{p-2}, \qquad \delta = p-1$$

The characteristic dimensionality d_c is that the value of d which validates Josephson's scaling law for the Landau exponents:

$$vd_c = 2 - \alpha = 2\beta + \gamma$$

This gives

$$d_c = \frac{p\sigma}{p-2} \qquad (2.27)$$

which can be rewritten more symmetrically as

$$\left(\frac{2d_c}{\sigma} - 2\right)(p-2) = 4 \qquad (2.28)$$

Finally, the Ginzburg criterion, linking the width of the critical region to the coefficient B_p, can be written

$$(B_p)^{2/(p-2)} \sim (\Delta T)^{(d_c - d)/\sigma} \qquad (2.29)$$

One can now see how the exponents in the different contributions to the free-energy density govern the classical values of the critical exponents, the characteristic dimensionality, and the Ginzburg criterion.

Setting $p = 4$ and $\sigma = 2$, one evidently recovers the expressions given in the preceding sections. For $p = 4$ and $\sigma < 2$, one obtains $d_c = 2\sigma$; in other words given any dimensionality d, there exists a critical value of σ, $\sigma = \dfrac{d}{2}$, below which the forces have a range long enough to validate Landau theory. In Chapter 10 we elaborate this question further, along with crossover effects in the presence of both short- and long-range forces.

Another interesting special case is $p = 6$ and $\sigma = 2$, which applies to the so-called tricritical points which have been mentioned several times already. We can give a brief definition of tricritical points as follows. There exist many systems where the order of a phase transition can be changed by varying a parameter, say the pressure; the tricritical point is the limiting case which separates first-order transitions (with a discontinuity in the order parameter) from second-order transitions (without discontinuity). In Landau theory, an expression like

$$F_L - F_0 = AM^2 + B_4 M^4 + B_6 M^6 + \cdots \tag{2.30}$$

yields a second-order transition for $B_4 > 0$, a first-order transition for $B_4 < 0$, (see Figure 2.2), and a tricritical point for $B_4 = 0$. Suppose now that B_4 is a function of, say, the pressure P; then by varying P we can eventually encounter a tricritical point. Chapter 12 discusses this more thoroughly, justifying incidentally the nomenclature 'tricritical'. Here we note only that the characteristic dimensionality is

$$d_c = 3 \tag{2.31}$$

Thus, as regards tricritical points, our three-dimensional world is precisely at

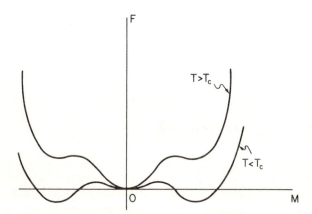

Figure 2.2. Plots of the function $F = F_L - F_0 = AM^2 + B_4 M^4 + B_6 M^6$ for two values of A, with B_4 negative and B_6 positive. In Landau theory, as A decreases there is a first-order transition when the two minima of F are equal; at that point the average magnetization changes discontinuously from zero to a finite value

the limit of validity of Landau theory; this limiting case entails logarithmic corrections to the theory, a not uncommon occurrence (see Chapters 5 and 13). Note that the classical tricritical exponents

$$\alpha = \tfrac{1}{2}, \qquad \beta = \tfrac{1}{4}, \qquad \delta = 5$$

differ from the ordinary classical exponents.

2.6 On the dimensionality n of the order parameter

Before discussing the effects of the dimensionality n, we summarize those of the space dimensionality d. To begin with, intuition suggests that the lower d, the more important the fluctuations; in one dimension ($d = 1$) any local fluctuation divides the system into two non-interacting parts, and the lower the dimensionality d, the fewer the paths around a local fluctuation. Accordingly, one expects mean-field theory to improve as the dimensionality rises; as regards the value of T_c this is what actually happens, with T_c increasing with d and tending asymptotically $(d \to \infty)$ to the mean-field value. The surprise in critical phenomena is that there exists a characteristic dimensionality making a sharp division between the two different regimes.

So far we have said little about the dimensionality n. In the foregoing sections (2.2 to 2.5), we considered order parameters with but one component; indeed both the Gaussian model and Landau theory decouple the different components from one another, and taking $n \neq 1$ would have made no difference to the calculation of the classical critical exponents. For non-classical critical exponents the situation is different, and one can try to foresee qualitatively the effects of varying n. Compare for instance the Ising model ($n = 1$) and the Heisenberg model ($n = 3$). In the Ising model the loss of long-range order implies spin-reversal, which is costly in energy; by contrast, the Heisenberg model needs only a very gradual rotation of the spin direction, which is the less expensive the more gradually it occurs. Thus, increasing n always favours fluctuations, tends to introduce new ones, and tends to lower the value of T_c. The present section aims to give a general survey of these effects of n.

Physical systems or models with different values of n

$n = 1$ includes the Ising model, magnetic systems with one axis of easy magnetization, and many cases where the order parameter is a genuine scalar, like liquid-to-gas transitions and order–disorder transitions in alloys.

$n = 2$ includes the XY model (Section 2.1) and magnetic systems with a plane of easy magnetization. In these cases it is convenient to distinguish between the amplitude and the phase of the order parameter; when spontaneous magnetization sets in, the symmetry it breaks is symmetry under rotations, i.e. a continuous symmetry, which entails that at low temperatures there exist infinitely many degenerate states obtainable from each other by rotating the magnetization. A good working definition of n is that there should exist $(n - 1)$ gauge variables and an $(n - 1)$-fold degenerate ground state, since one needs $(n - 1)$ independent continuous variables in order to specify one amongst the ensemble of all possible ground states. The superfluid systems (superconductors and Helium-4) have a complex order parameter, and $n = 2$. Other examples with $n = 2$, having an order parameter which is a density wave possessing amplitude and phase, are the smectic-A to nematic transitions of liquid crystals, and the Peierls transition (in absence of phase blockage by the lattice). The smectic-C to smectic-A transition, which also has $n = 2$, is fairly analogous to the XY model.

$n = 3$ is the Heisenberg model (see Section 2.1).

$n = 18, n = 6$. In the low-temperature A and B phases of superfluid Helium-3, the order parameter is a complex 3×3 matrix ($n = 18$), (Jones, Love, Moore, 1976). When dipolar interactions are taken into account, the order parameter of the 'planar' phase reduces to 3 complex components ($n = 6$).

Somewhat more generally we can list the following cases:

$n = d,$ where the order parameter is a vector (as in the Heisenberg model for $d = 3$).

$n = \dfrac{d(d - 1)}{2},$ where the order parameter is an antisymmetric tensor.

$n = \dfrac{d(d + 1)}{2} - 1,$ where the order parameter is a symmetric traceless tensor (as for biaxial nematics).

It is instructive to consider also the limiting case $n \to \infty$, regarded as a model; partly because it is soluble and its solution most informative, and partly because there is no reason to exclude the possibility of high physical values of n. The first exact solution for $n \to \infty$ was obtained in a classical Heisenberg model with an n-component order parameter; the result then turned out to be isomorphic to those already familiar in two other models, namely in the spherical model, and in the condensation of an ideal Bose gas at constant volume. The latter is a trivially soluble problem for a gas of non-interacting bosons, the transition being due simply to the Bose statistics. The spherical is related to the Gaussian model; it consists of a lattice of N spins, each having one component with a Gaussian probability distribution, but, and this is the difference from the Gaussian model, subject to the constraint

$$\sum_{i=1}^{N} M_i^2 = N \qquad (2.32)$$

This constraint is implemented by introducing a Lagrange multiplier which links the spherical and the Gaussian models through a Legendre transformation, analogous to that which connects the condensations of an ideal Bose gas at constant volume and at constant pressure respectively. Indeed, the isomorphism between the spherical model and ideal Bose condensation at constant volume is mirrored by an isomorphism between the Gaussian model and ideal Bose condensation at constant pressure.

These examples show that the value of n for a given model is not always obvious in advance; thus in the spherical model one starts with a model of one-component spins and winds up with an effectively infinite value of n. The dimensionality n is evidently a much more subtle quantity than the space dimensionality d; to discover what value of n is appropriate to a given model or physical system is often an important step in elucidating the whole problem.

Reverting to the case $n \to \infty$, note that in addition to taking the limit one can also expand in powers of $1/n$.

Simply by remarking that the excluded-volume problem for polymers corresponds to $n = 0$, de Gennes (1972) achieved an exceptionally elegant synthesis; at one stroke the excluded-volume problem is linked to phase transitions, while the physically significant range of n is enlarged. The value $n = 0$ should of course be understood in the sense of an analytic continuation, analogous in this respect to the space-dimensionality d when considered as a real variable; to the use of complex numbers in a whole host of problems (electric circuits, etc.); and to the introduction of non-integral values of the angular momentum ℓ in Regge theory, which is actually a very apt example, since for one-dimensional systems there is a close connection between analytic continuation in n on the one hand, and on the other hand continuation in ℓ for an equivalent quantum problem.

Once the case $n = 0$ is understood, there arises the further question about negative values of n. Here it turns out that for $n = -2$ the critical exponents assume extremely simple values (Balian and Toulouse, 1973): indeed there exists a close though only partial isomorphism between the case $n = -2$ on the one hand, and on the other hand the condensation of an ideal Bose gas at constant pressure.

At this stage we must summarize and rearrange our results.

In the standard case of ordinary critical points in spaces of dimensionality $2 < d < 4$, it seems that between $n = -2$ and $n = \infty$, the critical exponents are continuous function of n. Exact results are known for the limiting cases $n = -2$ and $n = \infty$, between which there occurs the case $n = 0$ of the excluded-volume problem for polymers (selfavoiding random walk), and the cases $n = 1, 2, 3, \ldots$ of normal phase transitions.

Accordingly it is convenient to design a table to serve as a kind of Mendeleff Periodic Table of simple systems, such that the position in the table of a model or of a physical system depends on the values of the two dimensionalities n and d; (see Figure 6.1). Systems represented by the same point are equivalent ('isotopes' of each other) as regards their static critical properties. As with the periodic table, there are very valuable possibilities of prediction and of interpolation between neighbouring points (see Chapter 6). Especially interesting in this respect are the expansions in powers of $\varepsilon = 4 - d$ or of $1/n$.

The case $n \to \infty$

Let the local free-energy density be given by

$$F_{\mathrm{L}} - F_0 = \frac{r_0}{2} \sum_{i=1}^{n} M_i^2 + \frac{u_0}{4n} \left(\sum_{i=1}^{n} M_i^2 \right)^2 + \frac{1}{2} \sum_{i=1}^{n} (\nabla M_i)^2 \qquad (2.33)$$

Note the denominator n in the coefficient of the quartic term; without it the problem would lack a finite limit as $n \to \infty$.

In the limit $n \to \infty$ the Hartree approximation becomes exact; (it amounts to a selfconsistent decoupling of the components M_i of the order parameter). We shall start by explaining briefly how the Hartree approximation is implemented when there are several components, and shall then demonstrate its validity in the limit $n \to \infty$.

The Hartree approximation consists in rewriting the quartic terms as

$$M_i^2 M_j^2 \to M_i^2 \langle M_j^2 \rangle + \langle M_i^2 \rangle M_j^2,$$

where $\langle A \rangle$ denotes the average value of A. This procedure decouples the fluctuations in the different components. At high temperatures, $T > T_c$, one has

$$\sum_i \langle M_i^2 \rangle = n \langle M_i^2 \rangle$$

all components playing an equivalent role. One can then rewrite F_L in the form

$$F_L - F_0 = \frac{r}{2} \sum_{i=1}^{n} M_i^2 + \frac{1}{2} \sum_{i=1}^{n} (\nabla M_i)^2 \tag{2.34}$$

which is typical of the Gaussian model (2.5), but with the additional selfconsistency condition

$$r = r_0 + u_0 \langle M_i^2 \rangle$$

Now introduce the correlation function $\Gamma(R) = \langle M_i(0)M_i(R) \rangle$. The self-consistency condition becomes

$$r = r_0 + u_0 . \Gamma(R = 0) = r_0 + u_0 \int \Gamma(k)\, dk$$

where $\Gamma(k)$ is the Fourier transform of $\Gamma(R)$, which in view of (2.34) has the Gaussian form

$$\Gamma(k) = \frac{T}{r + k^2} \tag{2.35}$$

This then leads to

$$r = r_0 + u_0 T \int \frac{dk}{r + k^2} \tag{2.36}$$

This last equation for r introduces the convention

$$r_0 = T - T_c$$

whence, by virtue of the definition of the exponent γ, one has in the vicinity of T_c

$$r = (T - T_c)^\gamma = (\Delta T)^\gamma$$

Notice that the proportionality constants have been dropped since at the moment we are interested only in the critical exponents. Now the transition temperature T_c

is obtained from (2.36) by setting $r = 0$:

$$0 = T_c - T_0 + u_0 T_c \int_0^\Lambda k^{d-3} dk \qquad (2.37)$$

Evidently one has $T_c < T_0$, for any dimensionality; for $d \leqslant 2$ the integral has an infrared divergence which reduces T_c to zero. For $d > 2$ it is convenient to subtract (2.37) from (2.36), whence

$$r = \Delta T \left[1 + u_0 \int_0^\Lambda k^{d-3} dk \right] - u_0 . T . r \int_0^\Lambda \frac{k^{d-1}}{k^2(k^2 + r)} dk \qquad (2.38)$$

For $d > 4$, the integral in the last term on the right converges when $r = 0$; in that case one has, for small r,

$$r \sim \Delta T$$

By contrast, for $2 < d < 4$, the integral is a singular function of r and the dominant term is

$$r \sim (\Delta T)^{2/(d-2)}$$

More generally we can write r as

$$r \sim \Delta T[1 + \Delta T + \cdots] + (\Delta T)^{2/(d-2)}[1 + \Delta T + \cdots]$$

and dominance passes from one term to the other at $d = 4$, where the two exponents are equal.

The critical exponents then follow from the equations (2.34), (2.35) and (2.38), one finds

for $d > 4$: $\qquad \eta = 0, \quad \gamma = 1, \qquad v = \frac{1}{2}, \qquad \alpha = 0$

for $2 < d < 4$: $\qquad \eta = 0, \quad \gamma = \frac{2}{d-2}, \quad v = \frac{1}{d-2}, \quad \alpha = \frac{4-d}{d-2}$

It is straightforward to apply the Hartree method to determine the equation of state, in spite of the somewhat pathological features of the low-temperature phase. In the presence of an external field H conjugate to the order parameter, one must distinguish between longitudinal and transverse components. One finds

$$H = r_L . M_L + \frac{u_0}{n} M_L^3$$

plus the selfconsistency condition

$$r_L = r_0 + u_0 T \int \frac{d\mathbf{k}}{r_T + k^2}, \quad \text{where } r_T = \frac{H}{M_L}$$

An argument exactly like the above leads to the exponents β and δ:

for $d < 2$: $\beta = \frac{1}{2}$, (β is well-defined though the magnetization is only of order $1/n$);

for $d > 4$: $\qquad \delta = 3$; $\qquad$ for $2 < d < 4$: $\qquad \delta = \frac{d+2}{d-2}$.

Also of interest is the longitudinal susceptibility in the low-temperature phase:

$$\chi_L = \frac{M_L}{H}\bigg|_{H\to 0} \sim \left(\frac{1}{H}\right)^{(4-d)/2} \tag{2.39}$$

Accordingly, the susceptibility is strictly infinite throughout the domain where the low-temperature phase exists; this is a general phenomenon, due to the existence of a gauge variable, and we shall return to it in Section 6.3.

It remains to justify the Hartree approximation in the limit $n \to \infty$. The most direct way is to write down a perturbation expansion in powers of the coupling constant $\frac{u_0}{n}$; in the limit $n \to \infty$, many terms become negligible, those that remain being those of the Hartree approximation, which in diagrammatic terms sums the tree diagrams for the self-energy. One can also without too much trouble re-sum the subdominant contributions of order $1/n$, which corresponds to the random-phase approximation (RPA). This hierarchy of approximations, RPA succeeding Hartree, is well known in the n-body problem; for the interacting electron gas it corresponds to an expansion in powers of the parameter r_s which governs the electron density, while here it amounts to an expansion in powers of $1/n$.

Here we must clarify an important point. At the start of this chapter we talked about mean-field theories, isomorphic to Landau theory, meaning theories which consider the fluctuations of a given spin, say, as taking place in the averaged field due to its *neighbours*. Such mean-field theories become exact in the limit where the number of neighbours tends to infinity; this can happen, in particular, if the spatial dimensionality is infinite, or if the interactions decrease slowly enough with distance. The two terms, mean-field approximation and Hartree approximation, are often used interchangeably. But in this book we distinguish between them, because the Hartree approximation to the coupling between different components, as described above, considers the fluctuations of one local component as taking place in the averaged field of the other *components*; this involves an approximation of a profoundly different physical nature, clearly reflected in the fact that for $2 < d < 4$ the values of the critical exponents are quite different in the two cases. Therefore we shall adhere to our own terminology of mean-field approximation (as between neighbours) and Hartree approximation (as between components).

Thus it is clear that the role which in mean-field theory belongs to the number of neighbours is assigned in Hartree theory to the number of components, and it is hardly surprising that the Hartree approximation improves as n increases.

One can go still further. By changing variables $M_i \to M_i\sqrt{n}$ in (2.33), the partition function can eventually be rewritten in the form

$$Z \sim \int \sum_{i=1}^{n} \mathscr{D}M_i(\mathbf{x}) \exp\left[-n \int g(M_i, \nabla M_i)\,\mathrm{d}\mathbf{x}\right]$$

where $n \to \infty$. Ordinary integrals of this type are evaluated by the saddle-point method, after determining the minimum of the function appearing in the exponent;

(compare to the method of stationary phase, and to the JWKB-type semiclassical methods in quantum mechanics, where $\hbar$ is small and $\frac{1}{\hbar}$ large). Functional integrals are treated in the same way; it is thus, for instance, that in superconductivity the BCS solution of the BCS Hamiltonian is shown to be exact in the limit of infinite volume.

Arguments like these aim at an intuitive understanding, but to carry out the calculations to order $1/n$ or higher, it is convenient to exploit diagrammatic techniques (see the review article by Ma, 1973).

The case $n = -2$

The case $n = -2$ simplifies quite remarkably: the free energy and the correlation function in zero field are Gaussian, as if the quartic term were absent. Moreover, this holds for any value of the space dimensionality d. Consequently the fluctuations fail to shift T_c, one has $T_c = T_0$, and Gaussian values are observed for the following exponents

$$\gamma = 1, \qquad v = \tfrac{1}{2}, \qquad \eta = 0$$

$$\alpha = \frac{4 - d}{2}, \qquad (d < 4)$$

The resemblance to the Gaussian model is not complete: thus, the free energy in an external field and the higher-order correlation functions are non-trivial; by contrast to the Gaussian model, the low-temperature phase is well defined; and, though no direct calculations have been performed for arbitrary d, there is every reason to believe, from exact results for $d = 1$ and from expansions in powers of $(4 - d)$ for arbitrary n (see Chapter 5), that the scaling laws are obeyed, which would entail

$$\beta = \frac{d - 2}{4}, \qquad \delta = \frac{d + 2}{d - 2}, \qquad (d < 4)$$

As in the general case of arbitrary n, there exists therefore a critical region with width determined by the coefficient u_0 of the quartic term (compare to the Ginzburg criterion); what is remarkable and special to the case $n = -2$ is the fact that the critical exponents have the same values inside and outside the critical region. Moreover, in zero field the free energy $(T > T_0)$ and the correlation function are completely unaffected by the existence of u_0 and of a critical region; by contrast, and as shown by the exact solution in one dimension $(d = 1)$, the higher order correlations have differential functional forms inside and outside the critical region.

The next task is to define exactly how n is to be continued to arbitrary real values. One possibility is to consider the perturbation series in powers of u_0; the various diagrams representing the terms of this series enter with weight factors involving polynomials in n, and one could decide to treat these n as a real variable. Another equivalent method uses auxiliary fields in the way customary in the theory of functional integration; this procedure yields a compact expression featuring n as a variable whose value is arbitrary. Let us consider the local free-energy density in an external field:

$$F_L - F_0 = \frac{1}{2}r_0 \sum_{i=1}^{n} M_i^2(\mathbf{x}) + u_0 \left(\sum_{i=1}^{n} M_i^2(\mathbf{x}) \right)^2 + \frac{1}{2}\sum_i (\nabla M_i)^2 - H \cdot M_1(\mathbf{x})$$

along with the resulting partition function

$$\frac{Z}{Z_0} = \int \prod_{i=1}^{n} \mathcal{D} M_i(\mathbf{x}) \exp\left(-\beta \int d\mathbf{x}[F_L - F_0] \right)$$

The method of auxiliary fields consists in introducing the field variable

$$P(\mathbf{x}) = \sqrt{u_0} \sum_{i=1}^{n} M_i^2(\mathbf{x}),$$

$$Z \sim \int \mathcal{D} P(x) \prod_{i=1}^{n} \mathcal{D} M_i(\mathbf{x})$$

$$\times \exp\left\{ -\beta \int d\mathbf{x} \left[\frac{1}{2}r_0 \sum_{i=1}^{n} M_i^2(\mathbf{x}) + P^2(x) + \frac{1}{2}\sum_i (\nabla M_i)^2 - H \cdot M_1(\mathbf{x}) \right] \right.$$

$$\left. \times \delta\left[P(\mathbf{x}) - \sqrt{u_0} \sum_{i=1}^{n} M_i^2(\mathbf{x}) \right] \right\}$$

and then replacing the δ-function by its integral representation $\delta(y) \sim \int \exp(i\lambda y)\, dy$; this leads to

$$Z \sim \int \mathcal{D} \lambda(\mathbf{x}) \int \mathcal{D} P(\mathbf{x}) \exp[i\lambda(\mathbf{x})P(\mathbf{x}) - \beta P^2(\mathbf{x})] \int \prod_{i=1}^{n} \mathcal{D} M_i(\mathbf{x})$$

$$\times \exp\left\{ -\frac{\beta}{2} \int d\mathbf{x} \left[\left(r_0 - \frac{2i\lambda(\mathbf{x})\sqrt{u_0}}{\beta} \right) \sum_i M_i^2(\mathbf{x}) + \sum_i (\nabla M_i)^2 - 2HM_i(\mathbf{x}) \right] \right\}$$

One can see that manipulation of the quartic term has factorized the integrations over the different components; there remain only Gaussian integrals of the type

$$\int \mathcal{D} M(x) \exp\left[-\frac{1}{2} \int d\mathbf{x}\, d\mathbf{x}'\, M(\mathbf{x})A(\mathbf{x}, \mathbf{x}')M(\mathbf{x}') + \int d\mathbf{x}\, C(\mathbf{x})M(\mathbf{x}) \right]$$

$$\sim (\det A)^{-(1/2)} \exp\left[\frac{1}{2} \int d\mathbf{x}\, d\mathbf{x}'\, C(\mathbf{x})A^{-1}(\mathbf{x}, \mathbf{x}')C(\mathbf{x}') \right] \qquad (2.40)$$

Here, the matrix A is a functional of $\lambda(x)$. Finally one obtains

$$Z \sim \int \mathcal{D} \lambda(\mathbf{x}) \exp\left(-\frac{\lambda^2(x)}{4\beta} + \frac{g\{\lambda\}}{2}H^2 \right) [\det A\{\lambda\}]^{-(n/2)} \qquad (2.41)$$

where $g\{\lambda\} = \beta^2 \int d\mathbf{x}\, d\mathbf{x}'\, A^{-1}(\mathbf{x}, \mathbf{x}', \{\lambda\})$. Every component contributes a power

$(-1/2)$ of the determinant; for n components its index is $(-n/2)$, and this is the only place where n occurs. One is then in a position to assign any arbitrary value to n in the expression (2.41).

For $n = -2$, the determinant has index 1. Since $\lambda(\mathbf{x})$ enters linearly into just one diagonal element of the matrix A, the determinant itself is linear in $\lambda(\mathbf{x})$; in zero field this linear term cannot contribute because the integrand is then odd in $\lambda(\mathbf{x})$; but u_0 enters only through the product $\sqrt{u_0}\,\lambda(\mathbf{x})$; hence the free energy in zero field is the same as it would be in the absence of u_0. It is clear that this argument fails when the field is non-zero, because the coefficient of H^2 introduces an additional dependence on $\lambda(\mathbf{x})$.

Accordingly, in our (n, d) tabulation the row $(n = -2, d$ arbitrary) is known. It is tempting to use this line as a basis for expansions in powers of $(n + 2)$, which with luck would complement the expansions in powers of $(4 - d)$ and of $1/n$. This programme has not yet been implemented.

So far we have considered only the standard case of ordinary critical points and short-range forces, i.e. $p = 4$ and $\sigma = 2$ in the notation of Section 2.5. The generalization to long-range forces, $(0 < \sigma < 2)$, is immediate. For tricritical points, $(p = 6)$, one obtains the amusing result that $n = -4$ is also a special case, giving the same exponents as $n = -2$ (Fisher, 1973).

Recall finally that the condensation of an ideal Bose gas at constant pressure provides a model which in some sense realizes the case $n = -2$ (Lacour-Gayet and Toulouse, 1974).

The excluded-volume problem for polymers and the case $n = 0$

Consider a chain laid out with its links arranged along the bonds between the sites of a lattice. One is interested in the statistics of such chains in the limit where the number N of links becomes large. This constitutes a random-walk problem, and also a model for studying the configurations of macromolecules (polymers).

In the absence of interactions between links, or in other words if the chain can cross a given lattice site repeatedly without impediment, one has the problem of an unconstrained random walk; this is the case of Brownian motion, which plays the same role here as the Gaussian model plays in phase transitions. A more realistic model for polymers introduces repulsive interactions; in the special case of a local infinitely repulsive interaction, one has the selfavoiding random walk problem.

Consider the number of chains (each with N links) having extension R, where R is defined as the distance between the two ends of the chain; this number defines a function $\Gamma_N(R)$ and its Fourier transform $\Gamma_N(k)$. It proves useful to introduce a kind of chemical potential P by defining

$$\Gamma(P, k) = \sum_N \exp(-NP)\Gamma_N(k)$$

This is the formulation which yields the closest analogy to the correlation function in the theory of phase transitions, P here playing the role of temperature. There exists a critical value P_c for which $\Gamma(P_c, k = 0)$ diverges. One then defines exponents as follows: γ by $\Gamma(P, k = 0) \sim \left(\dfrac{1}{P - P_{\mathrm{c}}}\right)^{\gamma}$; the exponent η by $\Gamma(P_{\mathrm{c}}, k) \sim \dfrac{1}{k^{2-\eta}}$, and so on. The most useful definition of v hinges on

$$\langle R^2 \rangle \sim N^{2v}, \qquad N \text{ large}, \tag{2.42}$$

which is consistent with the earlier definitions, as one can easily check.

For Brownian motion, i.e. for the unconstrained random walk, $\Gamma_N(R)$ and hence $\Gamma_N(k)$ are obtained by induction on N:

$$\Gamma_N(k) \sim z^N \exp\left(-\frac{N^2 k^2 a^2}{z}\right) \quad \text{for } k \text{ small},$$

where a is the length of each link and z the number of nearest neighbours of each lattice site.

This yields

$$\Gamma_N(R) \sim \exp\left(-\frac{R^2}{2\langle R^2 \rangle}\right) \quad \text{where} \quad \langle R^2 \rangle \sim Na^2 \text{ for } N \text{ large};$$

$$\text{whence} \quad v = \tfrac{1}{2}$$

and further

$$\Gamma(P, k) \sim \frac{1}{1 - z \exp\left(-P - \dfrac{k^2 a^2}{z}\right)}$$

which determines the critical value P_{c} through $e^{P_{\mathrm{c}}} = z$, and the exponents γ and η through

$$\Gamma(P \to P_{\mathrm{c}}, k = 0) \sim \frac{1}{P - P_{\mathrm{c}}}, \quad \text{whence} \quad \gamma = 1$$

$$\Gamma(P = P_{\mathrm{c}}, k \to 0) \sim \frac{1}{k^2}, \qquad \text{whence} \quad \eta = 0$$

Thus one obtains the classical (Gaussian) values for the exponents γ, η and v.

For the selfavoiding random walk (excluded-volume problem), numerical calculations allow one to form an altogether different picture. Here it seems that

$$\Gamma_N(k = 0) \sim \mu^N N^{\gamma - 1}, \quad \text{for} \quad N \text{ large},$$

instead of z^N as for Brownian motion. The constant μ is called the connectivity and determines the critical value P_{c} through

$$e^{P_{\mathrm{c}}} = \mu$$

Like T_{c} in phase-transition theory, the magnitude of μ has a very low degree of universality. Calculating $\Gamma(P, k = 0)$ from the expression for $\Gamma_N(k = 0)$ yields

$$\Gamma(P, k = 0) \sim \int dN\, e^{-NP} \mu^N N^{\gamma - 1} = \int dN \cdot N^{\gamma - 1} e^{(P_c - P)N}$$

$$\sim \left(\frac{1}{P - P_c}\right)^{\gamma}, \quad \text{for} \quad P \gtrsim P_c,$$

which motivates the notation chosen for the exponent of N in (2.43). Thus the value of γ is obtained from a numerical calculation of $\Gamma_N(k = 0)$. The value of the exponent v is obtained numerically, from $\langle R^2 \rangle$ using (2.42).

The numerical results are as follows (Shante and Kirkpatrick, 1971):

three dimensions $(d = 3)$: $2v \approx \frac{6}{5}$, $\gamma \approx \frac{7}{6}$
two dimensions $(d = 2)$: $2v \approx \frac{3}{2}$, $\gamma \approx \frac{4}{3}$

The numerical values are displayed as rational fractions, in deference to the Daltonian prejudice (so named in jest) that the exact results should assume such a form. This does in fact happen in several soluble models, for instance the Ising model in two dimensions; however, certain features of the expansions in powers of $(4 - d)$ and of $1/n$ militate against the prejudice.

Before discussing the connection with $n = 0$, we should mention that in earlier days the excluded-volume problem for polymers had elicited a whole host of theories. Amongst these we name two that are more or less selfconsistent, Flory's which predicted $v = \dfrac{3}{d + 2}$, and des Cloizeaux's which predicted $v = \dfrac{2}{d}$.

In order to take local repulsion into account, one could envisage a perturbation expansion of $\Gamma_N(R)$, starting from its Brownian-motion value, and subtracting from it successively the number of configurations with one self-intersection, with two intersections, and so on. Unfortunately it then transpires that for $d = 3$ this is an expansion in powers of $u_0 N^{1/2}$, which is very large. It was at this stage that de Gennes pointed out the rigorous isomorphism between such a perturbation expansion of $\Gamma_N(R)$ and the expansion of the correlation function $\Gamma(T, R)$ in the Ginzburg–Landau theory for $n = 0$. This immediately neutralizes the fact that the expansion coefficient is large, because the critical exponents are independent of the coupling constant u_0. In particular, the exponents γ, v, and η remain unchanged for any repulsive short-range interaction; they are observable for chains of greater than a certain critical length, whose magnitude admittedly does depend on the coupling strength. Comparing the values of the exponent v when $n = 0$ to the predictions of Flory and des Cloizeaux, we see that they agree for $d = 4, (v = \frac{1}{2})$, but disagree already to first order in $(4 - d)$. The good agreement between Flory's prediction and the numerical calculations for $d = 3$ and $d = 2$ appears to be coincidental.

The connection between the case $n = 0$ without magnetic field, and the behaviour of an isolated polymer, can be extended to non-zero magnetic field; this allows one to study polymer solutions (des Cloizeaux, 1976). The two problems are connected as follows. Let ρ_p, ρ, and π denote, respectively, the number of polymers per unit volume, the number of monomers per unit volume, and the osmotic pressure. If the number $N = \dfrac{\rho}{\rho_p}$ of monomers per chain is large, then studying the polymer solution amounts to considering the magnetic problem in presence of a magnetic field H near to the critical point, $T \to T_c$, and in the limit $n \to 0$. The following correspondence then exists between ρ, ρ_p, π, and $\Delta F = F(H) - F(0)$, where F is the free energy of the magnetic problem:

$$\rho_p = -H \frac{\partial}{\partial H} \Delta F$$

$$\rho + 2\rho_p = T \frac{\partial}{\partial T} \Delta F$$

$$\pi = -\Delta F$$

2.7 Conclusion

We conclude this chapter by noting, in a wider context, that the Ginzburg criterion identifies Landau theory as a limiting case within the general theory of critical phenomena. To recover an old theory as a special case of the new has become a characteristic of the physical sciences; witness thermodynamics as the 'thermodynamic limit' of statistical mechanics, Newtonian mechanics as the low-velocity limit of relativistic mechanics, and classical mechanics as the limit of quantum mechanics 'for small $\hbar$'. In this light, one could associate each of these limits with an appropriate Ginzburg criterion.

Appendix 2.1

As an exercise on the excluded-volume problem for polymers, we compare the probabilities of a return to the origin in unconstrained and constrained random walks.

The probability of a return to the origin, i.e. of tracing a closed polygon, is given by the ratio

$$\frac{\Gamma_N(R \to 0)}{\int \Gamma_N(R) \, d\mathbf{R}} = \frac{\Gamma_N(R \to 0)}{\Gamma_N(k \to 0)} \tag{2.44}$$

of the number of configurations having zero extension ($R = 0$) to the total number of all possible configurations; both are to be evaluated for the same number of links N, which is taken to be large.

The problem is an interesting challenge to intuition. One can see intuitively that the interaction between links will lead to two conflicting effects: there will be an expansion effect tending to push the far end of the chain away from the origin, and also an opposite caging effect tending to trap the end within the molecule.

For the Brownian unconstrained random walk one has

$$\Gamma_N(R) \sim \exp\left(-\frac{R^2}{\langle R^2 \rangle}\right), \quad \text{where} \quad \langle R^2 \rangle \sim N,$$

and the ratio becomes

$$\frac{\Gamma_N(R \to 0)}{\int \Gamma_N(R)\, d\mathbf{R}} \sim \left(\frac{1}{N}\right)^{d/2} \quad \text{for } N \text{ large}$$

In the presence of repulsive interactions one has (compare 2.18),

$$\Gamma_N(R \to 0) \sim \left(\frac{1}{N}\right)^{1-\alpha}$$

$$\int \Gamma_N(R)\, d\mathbf{R} = \Gamma_N(k \to 0) \sim \left(\frac{1}{N}\right)^{-\gamma}$$

and the ratio becomes

$$\frac{\Gamma_N(R \to 0)}{\int \Gamma_N(R)\, d\mathbf{R}} \sim \left(\frac{1}{N}\right)^{\gamma+1-\alpha} \quad \text{for } N \text{ large}$$

The values of this exponent, obtained numerically, are

for $d = 3$, approx. $23/12$
for $d = 2$, approx. $11/6$

in excellent agreement (in view of Josephson's scaling law) with the values for γ and α given in Section 2.6 above.

Thus we can see that for $d < 4$, the expansion effect dominates, the probability of a return to the origin being smaller with interactions than without.

What happens for $d > 4$? Noting the classical values $\gamma = 1$ and $\alpha = 0$, one would expect *a priori* that the ratio should depend on $1/N^2$, though one would need to make sure that the coefficient of this term does not vanish. If the coefficient is non-zero, then the probability of a return to the origin would be greater with interactions than without. If the coefficient is zero, then the decision hinges on the next term. Unfortunately, no numerical results are available as yet for these exotic dimensionalities.

Appendix 2.2

The percolation problem

Percolation presents a problem strikingly analogous to that of phase transitions. In both cases one defines critical exponents and scaling laws.

We start by formulating the percolation problem. Consider an infinite lattice, whose sites are, at random, either allowed or forbidden, with a probability p that any given site is allowed. If p is small one sees a sprinkling of isolated clusters each consisting of interconnected allowed sites and surrounded by a sea of forbidden sites. As p increases, it reaches a critical value p_c above which there appears an infinitely large cluster; one says then that percolation is taking place, meaning that one can cross the lattice by going successively from one allowed site to a neighbouring allowed site. As an alternative to such percolation by sites, one can define percolation by bonds, where it is the bonds between sites that are allowed or forbidden at random; the probability p_c depends on the type of percolation, but the 'critical' properties do not.

There exists a direct correspondence between the percolation problem and the problem of ferromagnetism. Fortuin and Kasteleyn (1972) have shown that percolation corresponds to the limit $s \rightarrow 1$ of the Potts model with s states (somewhat like the limit $n \rightarrow 0$ for polymers). This model is an extension of the Ising model. To every site of a lattice one assigns s states; to each bond between lattice sites one assigns an energy which can assume only one or the other of two values, namely one value when the two sites are in the same state, and another higher value when they are in different states. Then the correspondence between percolation and the ferromagnetic problem is as follows:

$$(p - p_c) \leftrightarrow (T_c - T)$$

(probability $M(p)$ that an allowed site belongs to an infinite cluster) $\leftrightarrow$ (spontaneous magnetization)
(average size $\chi(p)$ of finite clusters) $\leftrightarrow$ susceptibility
(average number $G(p)$ of clusters) $\leftrightarrow$ free energy

It is logical to define $M(p)$ to be the order parameter, in view of

$$M(p) = 0 \quad \text{for} \quad p < p_c$$

$$M(p) \neq 0 \quad \text{for} \quad p > p_c$$

The analogy between the average size of clusters and the free energy stresses the fact that in percolation there is strictly speaking no interaction energy, whence the quantity playing the role of free energy originates wholly from the entropy.

To explain what is meant by average values, we define the probability $s . G(p, s)$ that a site chosen at random belongs to a finite cluster of size s. Next, the average number of clusters of size s is defined as

$$G(p, s) = \frac{s . G(p, s)}{s}$$

where the division by s avoids counting the same cluster more than once. Then the average number of clusters having any size that is finite is

$$G(p) = \sum_s G(p, s)$$

Choose a site at random: either it is forbidden, or it is allowed and belongs to a cluster of finite size, or it is allowed and belongs to an infinite cluster. Hence one has

$$1 = (1 - p) + \sum_s sG(p, s) + pM(s)$$

Finally, the average size of finite clusters is defined by

$$\chi(p) = \sum_s s^2 G(p, s)$$

For greater insight into the analogy with ferromagnetism, define the Legendre transformation

$$G(p, \mu) = \sum_s \exp(-\mu s) G(p, s)$$

where μ is a kind of chemical potential. Then the analogy is

$$\mu \leftrightarrow H$$

where H is the magnetic field in the ferromagnetic problem. One can check that indeed $M(p)$ corresponds to $\left. \frac{\partial G}{\partial \mu}(p, \mu) \right|_{\mu \to 0}$ and $\chi(p)$ to $\left. \frac{\partial M}{\partial \mu}(p, \mu) \right|_{\mu \to 0}$

Hence one expects the power-law behaviour

$$M(p) \sim (p - p_c)^\beta$$

$$\chi(p) \sim \left(\frac{1}{p - p_c} \right)^\gamma$$

together with homogeneity properties for the singular part of $G(p, \mu)$, which in turn entail scaling laws between exponents. Such power-law behaviour has in fact been verified by numerical calculations. For the exponents β and γ the numerical results are as follows (Essam and Gwilym, 1971):

three dimensions $(d = 3)$: $\gamma \sim 1 \cdot 69 \pm 0 \cdot 05$, $(\beta + \gamma) \sim 2 \cdot 2 \pm 0 \cdot 3$
two dimensions $(d = 2)$: $\gamma \sim 2 \cdot 37 \pm 0 \cdot 03$, $(\beta + \gamma) \sim 2 \cdot 4 \pm 0 \cdot 2$

An exact solution has been given for percolation on a Bethe 'lattice', also called a Cailey tree. Such a tree is a system of indefinitely branching lines where no two branches ever intersect again; it cannot be inscribed into any genuine lattice having finite dimensionality, since in any such lattice some topological constraints would be unavoidable. It is generally accepted therefore that a Bethe lattice corresponds to an infinite dimensionality $(d \to \infty)$. Be this as it may, the following exponents result:

$$\beta = 1, \qquad \gamma = 1, \qquad \delta = 2, \qquad \alpha = -1$$

Notice that the characteristic dimensionality for which these exponents obey Josephson's law is $d_c = 6$.

The existence of a characteristic dimensionality $d_c = 6$ can be explained in terms of the correspondence between the percolation problem and the Potts model with s states in the limit $s \to 1$. A Landau–Ginzburg formulation of the Potts model contains terms of order M^3; and by substituting $\sigma = 2$ and $p = 3$ in (2.27) one finds indeed $d_c = 6$. Thus for $d \geqslant 6$ the critical exponents for percolation are the classical ones, the same as for a Bethe lattice. For $d < 6$, the non-classical exponents have been calculated in the vicinity of $d = 6$ by the renormalization-group method (Chapter 5), by expansions in powers of $\varepsilon = 6 - d$ carried to order ε^2 (Harris *et al.*, 1975).

General References

Stanley, H. E. (1971), *Introduction to Phase Transitions and to Critical Phenomena*, Clarendon Press (Oxford).
The series of volumes edited by Domb, C. and Green, M. S. (1972. . .), *Phase transitions and Critical Phenomena*, Academic Press.

References

Balian, R., Toulouse, G. (1973), *Phys. Rev. Letters*, **30,** 544.
des Cloizeaux, J. (1976), *J. de Physique*, **37,** C 1.255.
Domb, C. (1973), in *Collective Properties of Physical Systems*, Nobel Symposium XXIV, Academic Press.
Essam, J. W., Gwilym, K. M. (1971), *J. Phys. C.* L228.
Fisher, M. E. (1973), *Phys. Rev. Letters,* **30,** 679.
Fortuin, C. M., Kasteleyn, P. W. (1972), *Physica* **57,** 536.
de Gennes, P. G. (1972), *Phys. Letters,* **38A,** 339.
Harris, A. B., Lubensky, T. C., Holcomb, W. K., Dasgupta, C. (1975), *Phys. Rev. Letters,* **35,** 327.
Jones, D., Love, A., Moore, M. (1976), *J. Phys. C,* **9,** 743.
Lacour-Gayet, P., Toulouse, G. (1974), *J. de Physique*, **35,** 425.
Ma, S. K. (1973), *Rev. Mod. Phys.*, **45,** 589.
Shante, V. K. S., Kirkpatrick, S. (1971), *Adv. in Phys.*, **20,** 325.

CHAPTER 3
Homogeneity properties and scaling transformations

It was the homogeneity rules for critical phenomena which, in the middle sixties, provided the phenomenological basis for a theory to unify a domain threatening to diversify more and more. This approach put universality at the focus of both theoretical and experimental researches; it constituted an intermediate stage, when hypotheses were framed that were tb be confirmed a few years later by the application to critical phenomena of the theory of the renormalization group.

In the early sixties, the results from Onsanger's exact solution of the two-dimensional Ising model (1947) had provoked numerical studies of simple three-dimensional models. Several critical exponents had been estimated numerically and there was a feeling that they were not mutually independent. This led to the formulation of the four scaling laws, first as inequalities and then as equalities between exponents. There followed the homogeneity hypotheses, containing the scaling laws and implying further properties, which provoked many experimental and theoretical investigations.

The homogeneity properties characterizing a critical point and its neighbourhood will be introduced below rather like a set of Russian dolls contained one within the other, progressing from the smallest to the largest. Section 3.1 gives the homogeneity hypotheses in their raw and native form. Section 3.2 introduces the concept of scaling transformation. Finally Section 3.3 reformulates the homogeneity hypotheses as covariance properties under dilatation. Section 3.4 fits these considerations into the much wider context of scale invariance in physics generally.

3.1 Homogeneity rules

Thermodynamic functions and the equation of state

Near a critical point the Gibbs free energy $G(T, H)$ can be split into a regular part G_r and a singular part G_s. The homogeneity hypothesis asserts that G_s is a homogeneous function, of degree $2 - \alpha$, of the variables H and t^Δ,

$$\left(t = \left| \frac{T - T_c}{T_c} \right| \right),$$

so that it may be written

$$G_s \sim |t|^{2-\alpha} f^{\pm}\left(\frac{H}{|t|^\Delta} \right) \tag{3.1}$$

where $+(-)$ correspond to $t > 0$, $(t < 0)$. The critical exponent α was defined in Section 2.1 and governs the singularity in the specific heat. The 'gap exponent' Δ is not really a new one; we shall see later that it can be expressed as a function of the other exponents defined in Section 2.1. (The name gap exponent derives from the fact that, according to (3.1), successive derivatives of G with respect to H, at $H = 0$, obey power laws with exponents spaced by gaps Δ.)

The expression (3.1) is an asymptotic one which, strictly speaking, is satisfied only when t and H tend to zero. Away from this limit corrections are to be expected. The functions $f^{\pm}(x)$, defined for $0 < x < \infty$, are analytic in the neighbourhood of $x = 0$ and obey power laws as $x \to \infty$. Successive derivatives of G with respect to H behave asymptotically in the same way as G. The first two derivatives M and χ can be written as

$$M = -\frac{\partial G}{\partial H} \sim |t|^{2-\alpha-\Delta} g^{\pm}\left(\frac{H}{|t|^{\Delta}}\right)$$

(3.2)

$$\chi = -\frac{\partial^2 G}{\partial H^2} \sim |t|^{2-\alpha-2\Delta} h^{\pm}\left(\frac{H}{|t|^{\Delta}}\right)$$

The implications are readily derived.

Consider first a system in zero field $(H = 0)$ and with $t \neq 0$ but near the critical point. The quantities $f^{\pm}(0)$, $g^{\pm}(0)$, $h^{\pm}(0)$ are finite since $f^{\pm}(x)$ is assumed analytic in the neighbourhood of $x = 0$. For the critical behaviour of the specific heat one finds

$$C \sim -T\frac{\partial^2 G}{\partial T^2} \sim t^{-\alpha}$$

(3.3)

consistently with the definition of the exponent α in Section 2.1. The exponents β and γ, defined in the same section, then follow from α and Δ through

$$\beta = 2 - \alpha - \Delta$$

$$\gamma = -2 + \alpha + 2\Delta$$

(3.4)

Eliminating Δ from these equations one finds Rushbrooke's scaling law

$$\alpha + 2\beta + \gamma = 2$$

(3.5)

Next, consider the critical point itself, $t = 0$, $H \neq 0$; then $x = H/|t|^{\Delta}$ tends to infinity, and $g^{\pm}(x)$ must behave like $x^{1/\delta}$ in order to satisfy the relation $H \sim M^{\delta}$ between field and magnetization, as given in Section 2.1. Elimination of powers of t imposes $\beta = \Delta/\delta$, whence

$$\Delta = \beta\delta$$

(3.6)

Eliminating Δ and α from (3.4) and (3.6) leads one to Widom's scaling law, which connects the exponents β and γ defined for $T \neq T_c$ to the exponent δ defined at $T = T_c$:

$$\gamma = \beta(\delta - 1)$$

(3.7)

Accordingly, the first consequences of the homogeneity hypothesis for the Gibbs free energy are the Rushbrooke and the Widom scaling laws, which, between them, interrelate the four 'thermodynamic' critical exponents $\alpha, \beta, \gamma, \delta$. But the hypothesis implies still more: the equation of state $M = f(T, H)$ may be written in the homogeneous form $\dfrac{M}{|t|^\beta} = g^\pm\left(\dfrac{H}{|t|^{\beta\delta}}\right)$. In this form, the equation of state is represented by two curves g^+ and g^- for $t > 0$ and $t < 0$, respectively. But it is more elegant and more natural to consider $\overline{M} = \dfrac{M}{H^{1/\delta}}$ as a function of $\bar{t} = \dfrac{t}{H^{1/\beta\delta}}$, which makes it unnecessary to distinguish between $t > 0$ and $t < 0$. The equation of state is then described by a single curve as illustrated in Figure 3.1, which plots the reduced magnetization $\overline{M}$ against the reduced temperature $\bar{t}$ for the magnetic insulator $CrBr_3$. The points are far from spreading indiscriminately throughout the $(M, \bar{t})$ plane, as they would do if unconstrained by the homogeneity rule (3.2).

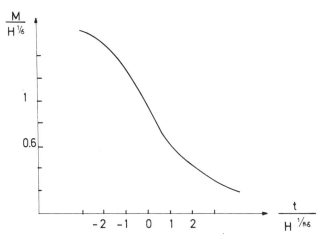

Figure 3.1. Plot of the scaling function in the equation of state $M/H^{1/\delta} = f(t/H^{1/\beta\delta})$, similar to that obtainable from experiments, for instance on $CrBr_3$; (experiment in fact yields points which sit on this curve). The scaling function $f(x)$ is normed by the conditions $f(0) = 1$ and $f(x) \sim (-x)^\beta$ as $x \to \infty$; these conditions correspond, respectively, to $H \sim M^\delta(t = 0)$ and to $M \sim (-t)^\beta$ for $(H = 0, t < 0)$

The hypothesis (3.1) allows one to display in very compact form results which at first sight seem unrelated. It holds in exactly soluble models like the spherical and the two-dimensional Ising models; and it is confirmed by numerical calculations on Ising models ($n = 1$), the XY model ($n = 2$), and Heisenberg models ($n = 3$) in two and three dimensions (Milosevic and Stanley, 1972). In these cases the critical exponents of successive derivatives of the free energy G with respect to the field H do indeed increase at every step by the same quantity $\Delta = \beta\delta$. Finally, the hypothesis is confirmed in the laboratory by experiments on many different systems;

54

confining ourselves to magnetic substances, we quote only $CrBr_3$, EuO, Pd_3Fe, Ni.

The fluctuation–dissipation theorem

So far we have been interested only in the thermodynamic functions G, M, χ. At the critical point these quantities display singular behaviour, attended by large fluctuations in the order parameter. The fluctuations lead to an instability signalled by a divergence in the susceptibility, which in general is connected with the appearance of a spontaneous magnetization. The relationship between fluctuations and the response function is prescribed by a general theorem on statistical mechanics, called the fluctuation–dissipation theorem.

Consider a classical system with the Hamiltonian $\mathscr{H}_0$, and apply a weak magnetic field $\mathbf{H}$, $\mathbf{H}$ being the field conjugate to the magnetization $\mathbf{M}$. Then the new Hamiltonian is

$$\mathscr{H} = \mathscr{H}_0 - \mathbf{H} \int d\mathbf{x}\, \mathbf{M}(\mathbf{x}) \tag{3.8}$$

The average value of the magnetization in unit volume is

$$\langle \mathbf{M} \rangle = \frac{1}{Z} \text{Trace} \frac{1}{\Omega} \int_\Omega d\mathbf{x}\, \mathbf{M}(\mathbf{x}) \exp(-\beta \mathscr{H}) \tag{3.9}$$

The susceptibility is defined as the response to this perturbation:

$$\chi = \frac{\partial \langle \mathbf{M} \rangle}{\partial \mathbf{H}} \tag{3.10}$$

Straightforward differentiation of (3.9) yields

$$\chi = \beta \int_\Omega d\mathbf{x} \{ \langle \mathbf{M}(0)\mathbf{M}(\mathbf{x}) \rangle - \langle \mathbf{M}(0) \rangle \langle \mathbf{M}(\mathbf{x}) \rangle \} \tag{3.11}$$

where we have appealed to translational invariance, which implies that in the thermodynamic limit of infinite volume the correlation function for two points depends only on the distance between them.

The theorem can be extended to the generalized susceptibility $\chi_{AB} = \dfrac{\partial \langle A \rangle}{\partial h_B}$, where h_B is the field conjugate to B. It is then written as

$$\chi_{AB} = \beta \int d\mathbf{x} \{ \langle A(0)B(\mathbf{x}) \rangle - \langle A(0) \rangle \langle B(\mathbf{x}) \rangle \} \tag{3.12}$$

In this way the specific heat can be expressed as an integral over the fluctuations of the energy density:

$$C \simeq - T\frac{\partial^2 G}{\partial T^2} \sim \beta^2 \int d\mathbf{x} \{ \langle \varepsilon(0)\varepsilon(\mathbf{x}) \rangle - \langle \varepsilon(0) \rangle \langle \varepsilon(\mathbf{x}) \rangle \} \tag{3.13}$$

These equations, though classical, apply to quantum systems as well, provided the observables in question commute with the Hamiltonian. Moreover, the classical limit is appropriate at high enough temperature, and as far as critical phenomena are concerned it applies whenever the critical temperature is non-zero. This point has already been made in our discussion of universality in Section 2.1.

Correlation functions

In Section 2.1 we defined the long-distance behaviour of the correlation function $\Gamma(R)$ at the critical point $T = T_c$, and also the divergence of the correlation length ξ in this neighbourhood. As regards Γ one adopts the homogeneity hypothesis

$$\Gamma(R, \Delta T) \sim \frac{1}{R^{d-2+\eta}} f_1 \left(\frac{R}{\xi} \right) \tag{3.14}$$

which clearly identifies ξ as the length-scale of the problem. Equation (3.14) describes the behaviour in the asymptotic region where $\Delta T = T - T_c$ tends to zero and where $R \gg R_0$, R_0 being the range of the forces. The Fourier transform $\Gamma(k, \Delta T)$ is directly measurable in diffraction experiments; it takes the homogeneous form

$$\Gamma(k, \Delta T) \sim \frac{1}{k^{2-\eta}} f_2(k\xi) \tag{3.15}$$

In the neighbourhood of the critical point the fluctuation–dissipation theorem can be written as

$$\chi = \beta \Gamma(k = 0, \Delta T) \sim t^{-\gamma} \tag{3.16}$$

Let us therefore consider the limit of (3.15) as k tends to zero; then $f_2(x)$ must vanish like $x^{2-\eta}$. For $\Gamma(k = 0, \Delta T)$ this implies the asymptotic behaviour

$$\Gamma(k = 0, \Delta T) \sim \xi^{2-\eta} \sim t^{-\nu(2-\eta)} \tag{3.17}$$

which by comparison to (3.16) leads to Fisher's scaling law

$$\gamma = \nu(2 - \eta) \tag{3.18}$$

This scaling law is a direct consequence of the homogeneity hypothesis for Γ taken together with the fluctuation–dissipation theorem; it links the thermodynamic exponent γ to the 'fluctuation' exponents ν and η.

The hypothesis (3.14) generalizes to correlation functions for several points, or rather to their cumulants† as follows

$$K_{AB...N}(\mathbf{x}_1 ... \mathbf{x}_n) \sim \frac{1}{R^{d-2+\eta(A...N)}} D_{AB...N} \left(\frac{\mathbf{x}_1}{\xi} ... , \frac{\mathbf{x}_n}{\xi} \right) \tag{3.19}$$

Here, R is a representative distance within the configuration $(\mathbf{x}_1 ... \mathbf{x}_n)$, and the

---

† Translator's note. The cumulants are often defined or referred to as the contribution of the connected diagrams.

cumulant $K_{AB...N}(\mathbf{x}_1...\mathbf{x}_n)$ is defined, by aid of auxiliary fields $h_A(\mathbf{x}_1)...h_N(\mathbf{x}_n)$ conjugate respectively to $A(\mathbf{x}_1)...N(\mathbf{x}_n)$, according to

$$K_{AB...N}(\mathbf{x}_1...\mathbf{x}_n) = \frac{\partial^n}{\partial \beta h_A(\mathbf{x}_1)...\partial \beta h_N(\mathbf{x}_n)} \log \text{Trace} (\exp - \beta \mathscr{H}) \bigg|_{\substack{h_A = 0 \\ h_N = 0}}$$

where

$$\mathscr{H} = \mathscr{H}_0 - \int h_A(\mathbf{x}_1) A(\mathbf{x}_1) \, d\mathbf{x}_1 - \cdots - \int h_N(\mathbf{x}_n) N(\mathbf{x}_n) \, d\mathbf{x}_n$$

In an external field H, the correlation function takes the homogeneous form

$$\Gamma(H, T) \sim \frac{1}{R^{d-2+\eta}} f\left(\frac{R}{\Delta T^{-\nu}}, \frac{H}{\Delta T^{\Delta}}\right) \tag{3.20}$$

This last formulation contains the preceding ones, and in particular the hypothesis (3.1) for the free energy; one can see this since a straightforward integration, as in the fluctuation–dissipation theorem, leads from Γ to χ and thence to M and to G. The scaling laws (3.5), (3.7) and (3.18) follow in turn, except for Josephson's which, alone, involves the space dimensionality d. In Sections 3.2 and 3.3 we shall reformulate the homogeneity properties in terms of concepts that we have seen emerging already, namely scaling transformations and covariance under dilatation. But before tackling scale transformations it is useful to generalize the homogeneity rules by taking additional parameters into account.

Generalized homogeneity, and crossover phenomena

Consider a system subject to fields, i.e. to intensive variables, in addition to the temperature, and to the magnetic field which is the conjugate of the order parameter. Use μ_i to denote these additional fields, which for an antiferromagnetic substance for instance could be the pressure or the uniform magnetic field. We assume that when $\mu_i = 0$, the critical behaviour obeys the homogeneity hypotheses (3.1) and (3.14) with critical exponents α, β, etc. When the fields μ_i are non-zero but weak, these hypotheses are generalized as follows. The correlation function takes the homogeneous form

$$\Gamma(\Delta T, H, \mu_i) \sim \frac{1}{R^{d-2+\eta}} D\left(\frac{R}{|t|^{-\nu}}, \frac{H}{|t|^{\Delta}}, \frac{g_i}{|t|^{\phi_i}}\right) \tag{3.21}$$

where the 'scaling fields' g_i are certain functions of the experimental fields. We associate a new exponent ϕ_i, called a crossover exponent, with every pair (g_i, O_i), where O_i, conventionally called an operator, is the density of the extensive variable conjugate to g_i. The expression (3.21) applies asymptotically in the limit where H, $|t|$, g_i all tend to zero. Then by virtue of the fluctuation–dissipation theorem the free energy also assumes a homogeneous form

$$G_s(\Delta T, H, \mu_i) \sim |t|^{2-\alpha} f\left(\frac{H}{|t|^{\Delta}}, \frac{g_i}{|t|^{\phi_i}}\right) \tag{3.22}$$

Accordingly, everything proceeds as if in the critical region the scaling field g_i were replaced by the reduced field $\bar{g}_i = g_i/|t|^{\phi_i}$. Depending on the sign of ϕ_i one distinguishes three situations that are qualitatively different:

i. ϕ_i negative. As the critical point is approached, ($|t| \to 0$), the reduced field $\bar{g}_i$ tends to zero. The leading singularity in the asymptotic behaviour is the same as if g_i were zero. Hence the operator O_i is irrelevant; but the Taylor expansion of the scaling function $f(x)$ about $x = 0$ corrects the dominant singular terms by addends proportional to $|t|^{2-\alpha+|\phi_i|}$.

ii. ϕ_i positive. As $|t|$ tends to zero it passes through three distinct regions, $|t|^{\phi_i} \gg g_i$, $|t|^{\phi_i} \sim g_i$ and $|t|^{\phi_i} \ll g_i$. In the first region, far from the critical point, $\bar{g}_i$ is small and everything proceeds as if g_i were zero; in a manner of speaking g_i is renormalized according to distance from the critical point. In the second region, centred on the crossover temperature $|t^*| \sim g_i^{1/\phi_i}$, the perturbation begins to make itself felt. Finally in the third region, closest to the critical point, the reduced field $\bar{g}_i$ becomes very large, which is simply a reminder that a perturbation treatment is no longer adequate. The true asymptotic behaviour as $|t| \to 0$ is certainly not the critical behaviour that would correspond to $\mu_i = g_i = 0$. Either there is no critical behaviour at all, as for instance in the presence of a magnetic field, which suppresses the transition, ($\Delta = \beta\delta$ if always positive can be thought of as a crossover exponent); or else the actual critical behaviour is quite different and depends on the nature of the operator O_i. Such crossover phenomena from one kind of behaviour to another reflect competition between two critical regimes; the first yields progressively to the second which is fully dominant at the critical point itself. It is reasonable to assume that there exists a single scaling function which describes the changeover from one critical regime to the other; this then implies that the critical temperature $t_c(g_i) = \dfrac{T_c - T_c(0)}{T_c(0)}$ shifts proportionally to $g_i^{|\psi_i|}$, where the shift exponent ψ_i is equal to ϕ_i.

iii. ϕ_i zero. This situation will be studied in detail in Chapter 13; it can have various rather pathological consequences, like logarithmic corrections to the power laws, or critical exponents varying continuously with the field g_i.

3.2 Scaling transformations

We start with a naïve dimensional analysis whose inadequacies will promptly become evident; and proceed to show how, taking a more general point of view, the homogeneity rules discussed in the preceding section can be derived from scaling transformations.

Dimensional analysis and canonical dimensions

The free-energy density G, or rather the product βG, has the dimensions

$$[G] = \frac{1}{[L^d]} \tag{3.23}$$

Next, let us assume that $\Gamma(R) = \langle M(0)M(R) \rangle$ obeys at $T = T_c$ a power law like that given in Section 2.1; this suggests that the order parameter M has dimensions

$$[M] = \frac{1}{[L^{(d-2+\eta)/2}]} \tag{3.24}$$

Then the fluctuation–dissipation theorem leads one to assign to the susceptibility χ the dimensions

$$[\chi] = \frac{[L^d]}{[L^{d-2+\eta}]} = [L^{2-\eta}] \tag{3.25}$$

Finally, the dimensions of H follow from those of M because the product HM has the same dimensions as G; thus

$$[H] = \frac{[L^{-d}]}{[M]} = \frac{1}{[L^{(d+2-\eta)/2}]} \tag{3.26}$$

If we now replace L by the characteristic length in the problem near the critical point, namely by the correlation length $\xi \sim t^{-\nu}$, then by appeal to the dimensions of G, M, χ and H, as given above, we recover the four scaling laws, including Josephson's; they follow in the form of expressions giving the thermodynamic exponents α, β, γ, δ in terms of the two exponents η and ν which refer to the correlation function.

At this stage we can try to continue the dimensional analysis, started in Section 2.4, of the Ginzburg–Landau form of the free energy. Its results were

$$[M] = \frac{1}{[L^{(d-2)/2}]}, \qquad [r_0] = \frac{1}{[L^2]}$$

where $r_0 \sim \Delta T$. Our previous analysis leading to the scaling laws remains valid, but now with the special values $\eta = 0$ and $\nu = \frac{1}{2}$. Thus one recovers the exponents of the Gaussian model; which is not very surprising, since this model contains only one length ξ_G, defined by $r_0 = 1/\xi_G^2$. But in the Ginzburg–Landau theory where u_0 is non-zero a second length enters through $\xi_1 = u_0^{1/d-4}$, and naïve dimensional analysis certainly ceases to apply. As long as ξ_G is smaller than ξ_1, i.e. not too near the critical point, the system is aware only of ξ_G and behaves as if u_0 were zero. But as soon as the two lengths become comparable one expects ξ to be given by some combination of both. As an illustration, consider the very special limiting case $n \to \infty$, which corresponds to the spherical model. One sees at once that ξ is a combination of ξ_1 and ξ_G,

$$\xi \sim (\xi_1^{d-4} \cdot \xi_G^2)^{1/(d-2)} \tag{3.27}$$

yielding the non-Gaussian value $\nu = \dfrac{1}{d-2}$ for the critical exponent ν; and it is obvious that dimensional analysis on its own cannot give such a result. Thus, the naïve dimensional analysis leading to the Gaussian values is in general incorrect.

For instance, the two-dimensional Ising model has $v = 1$ and $\eta = \frac{1}{4}$, quite unlike the Gaussian values $v = \frac{1}{2}$ and $\eta = 0$.

Such arguments lead one to denote as *canonical* the dimensions d_A arising from naïve dimensional analysis, $([A] = [L^{-d_A}])$, in contrast to the *anomalous* dimensions which are defined below by appeal to scaling transformations, and which result in non-Gaussian critical exponents. Table 3.1 shows the canonical dimensions d_A of several variables A.

Table 3.1. Canonical dimensions
d_A defined by $[A] = 1/[L^{d_A}]$

A	d_A
M	$\frac{1}{2}(d - 2)$
H	$\frac{1}{2}(d + 2)$
χ	-2
G	d
r_0	2
u_0	$4 - d$

Homogeneity rules, scaling transformations and anomalous dimensions

Once more we begin with the hypothesis that at the critical point the correlation function $\Gamma(R)$ falls off slowly and according to a power law. Consider now a scaling transformation, i.e. a dilatation in the unit of length,

$$a \to sa \tag{3.28}$$

distances being reduced accordingly:

$$R \to R' = R/s \tag{3.29}$$

Under this transformation the power law for Γ is *invariant* if we assume that the order parameter transforms as follows:

$$M \to M' = s^{(d - 2 + \eta)/2} M \tag{3.30}$$

This invariance property is characteristic of the critical point: one can say that under it there is complete equivalence between changing the distance R, and changing the scale of the order parameter. This comes about because in such cases the scale of the order parameter can be varied independently of the initial value of R, which would be impossible if for instance one had $\Gamma(R) \sim e^{-R/R^0}$.

Under the dilatation (3.28) the magnitudes of G, Γ, ΔT and H transform as follows:

$$G \to G' = s^d G$$

$$\Gamma \to \Gamma' = s^{d - 2 + \eta} \Gamma$$

$$\Delta T \to \Delta T' = s^{1/v} \Delta T \tag{3.31}$$

$$H \to H' = s^{y_h} H = s^{(d + 2 - \eta)/2} H$$

Here, the transformation for ΔT follows from that for ξ once one takes $\xi \sim \Delta T^{-\nu}$, and the transformation for H follows because the product MH transforms like G. Equations (3.31) imply the homogeneity rules

$$G(\Delta T, H) \sim s^{-d} G'(\Delta T, H) \sim s^{-d} G(s^{1/\nu} \Delta T, s^{y_h} H)$$

$$\Gamma(\Delta T, H, R) \sim s^{-(d-2+\eta)} \Gamma'(\Delta T, H, R) \sim s^{-(d-2+\eta)} \Gamma\left(s^{1/\nu} \Delta T, s^{y_h} H, \frac{R}{s}\right)$$

(3.32)

These are just the rules already given in Section (3.1) but in a different form. One form is obtained from the other by replacing $\Delta T s^{1/\nu}$ in (3.32) by 1, i.e. s by $\Delta T^{-\nu}$. However, (3.32) is more general than the formulation in Section 3.1, where all quantities were scaled to a temperature $|t|$.

We shall take any general quantity A to transform under dilatation according to the rule

$$A \rightarrow A' = s^{d_A} A,$$

which defines the *anomalous* dimension d_A of A.

3.3 Covariance under dilatation

The generalized homogeneity rules can be reformulated as invariance properties under a transformation (dilatation) which alters the scale of length,

$$\mathbf{x} \rightarrow \mathbf{x}' = \mathbf{x}/s,$$

and also renormalizes the scaling fields g_i and their conjugate operators $O_i(\mathbf{x})$ as follows:

$$g_i \rightarrow g_i' = s^{y_i} g_i \tag{3.33}$$
$$O_i(\mathbf{x}) \rightarrow O_i'(\mathbf{x}') = s^{x_i} O_i(\mathbf{x}) \tag{3.34}$$

In principle the local free-energy density $F_L(\mathbf{x})$ defined in Section 2.1 can be written as a sum of products $g_i O_i(\mathbf{x})$ of fields and operators where the field (e.g. the temperature) is an intensive variable and the conjugate operator (e.g. the energy) is the density of an extensive variable:

$$F_L(\mathbf{x}) = \sum_i g_i O_i(\mathbf{x}).$$

With every pair $(g_i, O_i(\mathbf{x}))$ we associate two exponents x_i and y_i, which are not independent. Under a scaling transformation the product $g_i O_i(\mathbf{x})$ transforms like the free-energy density G, whence we have

$$G \rightarrow G' = s^d G$$

$$g_i O_i(\mathbf{x}) \rightarrow g_i' O_i'(\mathbf{x}) = s^{x_i + y_i} g_i O_i(\mathbf{x})$$

and consequently

$$x_i + y_i = d \tag{3.35}$$

For instance, with the pair (energy and temperature) we associate the anomalous dimensions d_{ϕ^2} and $y_E = \dfrac{1}{\nu}$, where $d_{\phi^2} + y_E = d$; with the pair (magnetization and magnetic field) we associate the anomalous dimensions d_ϕ and $y_h = \dfrac{d + 2 - \eta}{2}$, where $d_\phi + y_h = d$. (The notations d_ϕ and d_{ϕ^2} derive from field theory, where the field variable is traditionally denoted by $\phi(x)$ instead of $M(x)$.) The quantities directly accessible to experiment are the exponents α, β, γ, etc; they have simple expressions in terms of the anomalous dimensions defined above.

The covariance properties under dilatation of G, Γ, g_i, g_j, and so on are reflected in the homogeneity rules already given in Section 3.2 for the special case where $g_i = \Delta T$ and $g_j = H$:

$$G(g_i, g_j, \ldots) = s^{-d} G(s^{y_i} g_i, s^{y_j} g_j, \ldots)$$

$$\Gamma(g_i, g_j, \ldots, R) = s^{-2d_\phi} \Gamma\left(s^{y_i} g_i, s^{y_j} g_j, \ldots, \frac{R}{s}\right) \tag{3.36}$$

For the cumulants $K_{O_1 \ldots O_N}(\mathbf{x}_1, \ldots, \mathbf{x}_N)$ we can write the generalized homogeneity rule

$$K_{O_1 \ldots O_N}(\mathbf{x}_1, \ldots, \mathbf{x}_N, g_i, g_j, \ldots)$$

$$= s^{-(d_1 + \cdots + d_N)} K_{O_1 \ldots O_N}\left(\frac{\mathbf{x}_1}{s}, \ldots, \frac{\mathbf{x}_N}{s}, s^{y_i} g_i, s^{y_j} g_j, \ldots\right) \tag{3.37}$$

At the critical point the correlation length ξ is infinite, and it remains infinite under scaling transformations. This invariance property imposes the criticality condition $g_i = 0$ on every field g_i having $y_i > 0$. The fields g_i having $y_i > 0$ tend to zero under continued rescaling. Near the critical point the relevant fields, those having $y_i > 0$, are in competition, which leads to the crossover behaviour described in Section 3.1, with crossover exponent $\phi_{ij} = y_i / y_j$.

The dimensions y_i thus serve to classify the scaling operators $O_i(\mathbf{x})$ according to their degree of irrelevance for $y_i < 0$, or their degree of relevance for $y_i > 0$. The concept of the relevance of an operator will prove especially useful from Chapter 7 onwards, when we move on from simple systems to study complex ones, where an additional perturbation, if relevant, leads to new kinds of critical behaviour.

3.4 Scale invariance in physics

Critical phenomena and the theoretical techniques used to study them have some remarkable features which make them interesting as paradigms. For this if for no other reason one would be ill-advised to assimilate either the phenomena or the methods too closely to others that physics has dealt with in the past.

Nevertheless, without losing sight of the characteristics and simplicities typical of these phenomena, it is still useful to pursue certain lines of thought somewhat further.

One very powerful concept involved is that of scale invariance; some properties are to be understood as invariance properties arising from a symmetry of the system, in this case a dilatation symmetry in space.

The underlying idea, that the laws of nature are independent of our choice of units, leads directly to the usual rules of dimensional analysis. Next we invoke the physical argument that certain types of phenomena, characterized by a definite scale of length, may cease to depend on dimensional parameters whose scale is very different. Systematic and wholly non-trivial applications of these two ideas can be found in the similarity rules for hydrodynamic flow, in the theory of turbulence (Kolmogorov) and in the theory of gaseous combustion (Landau and Lifshitz, 1959). Such considerations also play a central role in all work involving simulation on reduced-scale models.

Further, we have the mathematical definition of dilatation operations along the same lines as the more familiar operations of rotation and translation. Thus, if in a classical single-particle problem the Hamiltonian is dilatation-symmetric (which may be harder to detect than rotational or translational symmetry), then this gives rise to a certain constant of the motion (Jackiw, 1972). As with any other symmetry, once the invariance properties under dilatation are made explicit, they can shed much light on the solution of certain specific problems (involving one or many particles). Moreover, and still in analogy with other symmetries, our interest is not confined to cases where the symmetry is exact, but extends to cases where it is broken, either by a preferably small perturbation, or spontaneously.

From this point of view it seems interesting to consider an even more extended group of spatial symmetries. To the translations (d), rotations $\left(\dfrac{d(d-1)}{2}\right)$, and dilatations (1), we adjoin the special conformal transformations (d), (the number of group parameters (group dimensionality) is shown in brackets), and construct thus the group of conformal transformations conserving angles; its dimensionality D, in a space of d dimensions, is

$$D = \frac{d(d+3)}{2} + 1,$$

except when $d = 2$, in which case the conformal group has infinite dimensionality. The study of conformal invariance properties seems promising but has been too little developed as yet to be discussed here (Dubovik, 1973; Fisher, 1973).

The concept of anomalous dimensions, illustrated so remarkably by critical phenomena, has forerunners in certain model field theories. In the case of critical phenomena, the non-classical values of the critical exponents demonstrate clearly that the concept is relevant to physical reality. The reason why these exponents can be interpreted in terms of powers of length lies in the scaling transformations, which do not reduce to a naïve dilatation of all lengths, but only (and subtly) of those lengths that govern the physics.

The argument of this section will be extended and better founded in Chapter 14, where we shall discuss the concept of scale invariance in elementary-particle theory.

3.5 Conclusion

Later chapters will apply the theory of the renormalization group to critical phenomena in order to substantiate the preview given above; to show how to calculate the critical exponents and the scaling functions; and to elucidate the limits of validity of the asymptotic formulae by determining the correction terms. Besides, phenomenological scaling arguments of the kind we have used so far remain very useful tools for the preliminary unravelling of many problems; examples include dynamical critical phenomena (dynamic scaling) and finite-size effects.

General References

Reidel, E. K., Wegner, F. (1969), *Z. Physik*, **225,** 195.
Stanley, H. E. (1971), *Introduction to Phase Transitions and Critical Phenomena*, Clarendon Press (Oxford).
Widom, B. (1965), *J. Chem. Phys.*, **43,** 3898.

References

Dubovik, V. M. (1973), *Sov. Phys. Usp.*, **16,** 275.
Fisher, M. E. (1973), in *Collective Properties of Physical Systems*, Nobel Symposium XXIV, Academic Press.
Jackiw, R. (1972), *Physics Today*, **25,** 23.
Landau, L., Lifshitz, E. (1959), *Fluid mechanics*, Pergamon Press, London.
Milosevic, S., Stanley, H. E. (1972), *Phys. Rev.*, B **6,** 986, 1002.

CHAPTER 4
The renormalization group in the theory of critical phenomena

> '*Mathematics and physics are sciences whose yoke scholars never cease to feel; ultimately it should be abandoned, but meanwhile methods keep multiplying; the same spirit of improvement which discloses new vantage points in these fields also improves, through abridgment, our ways of learning about them, and provides new means of mastering the new domains which it discloses in the sciences.*'
>
> Fontenelle

In the introduction (Section 1.2) we sketched how the renormalization group aims to establish a family of correspondences. Our next task is to implement this programme for a specific problem, namely for critical phenomena treated within the framework of the Ginzburg–Landau formalism.

4.1 Definition of the group operations

The initial system, and trajectories in parameter space

The Ginzburg–Landau formulation described in Section 2.1 expresses the partition function Z as a functional integral over spatial fluctuations of the field variables $\mathbf{M}(\mathbf{x})$, (see Equation 2.1):

$$Z = \int \mathscr{D}\mathbf{M}(\mathbf{x}) \cdot \exp[-\beta \int F_L\{\mathbf{M}(\mathbf{x})\}\, d\mathbf{x}]$$

It proves useful to write Z as an integral over fluctuations $\mathbf{M}(\mathbf{k})$ with given wavenumbers $\mathbf{k}$. This amounts to changing the integration variables in the functional integral, and leads to the compact and basic expression

$$Z = \int \prod_{|\mathbf{k}| < 1/a} d\mathbf{M}(\mathbf{k}) \cdot \exp(-\mathscr{H}) \tag{4.1}$$

where two features call for comment. First, we have introduced a cutoff parameter $1/a$ at large wavenumbers, corresponding to a minimum distance a; for instance, on a lattice with lattice constant a there can be no fluctuations having wavenumbers greater than $1/a$. Second, $\mathscr{H}$ is defined by

$$\mathscr{H} = \beta F, \qquad F = \int F_L\{\mathbf{M}(\mathbf{x})\}\, d\mathbf{x} = \int F\{\mathbf{M}(\mathbf{k})\}\, d\mathbf{k}$$

Hence, strictly speaking, $\mathscr{H}$ is not a Hamiltonian. This is a convenient notation with the further advantage that it avoids any misleading impression that temperature plays a special role. In the following, whenever we speak of a system, it is to be understood that we mean a state of a system at a given temperature.

The initial system is defined in terms of its 'Hamiltonian' by the standard choice

$$\mathscr{H} = \mu_0 + \frac{1}{2} \int_{k<1/a} (r_0 + k^2)|M(k)|^2$$

$$+ u_0 \int_{k_1,k_2,k_3,k_4<1/a} M(k_1).M(k_2).M(k_3).M(k_4)\delta(k_1 + k_2 + k_3 + k_4) \tag{4.2}$$

where vectors are not specially labelled, in order to ease the notation. This choice of $\mathscr{H}$ is equivalent to the standard choice of F given in Equation (2.2). When $u_0 = 0$ one has the Gaussian model which is soluble exactly (see Section 2.2).

The choice of the $\mathbf{M}(\mathbf{k})$ as variables is motivated as follows. In the expression (4.1) for Z we proceed to carry out the integrations with respect to the $\mathbf{M}(\mathbf{k})$, but only over that part of the wavenumber range which lies between $1/a$ and $1/sa$. In this way a new 'Hamiltonian' $\mathscr{H}'$ is defined, symbolically, by

$$\exp(-\mathscr{H}') = \int \prod_{(1/sa)<k<1/a} d\mathbf{M}(\mathbf{k}) \exp(-\mathscr{H}), \quad s > 1 \tag{4.3}$$

The initial system is specified by the values of the parameters r_0 and u_0 in the 'Hamiltonian' $\mathscr{H}$, Equation (4.2). Note that we have chosen the coefficient of the k^2 term to be $\frac{1}{2}$. Then the initial system is further specified by the cutoff parameter $1/a$.

Next, we think of this system as one member of a large family of systems, which are specified by 'Hamiltonians' $\mathscr{H}$ containing couplings of all orders, coupling constants depending on the momenta, (i.e. non-local), and so on. However, the family is restricted to systems all having the same cutoff parameter $1/a$, and the same coefficient in the term involving $k^2 M^2$. In order that the 'primed' system above should belong to this family, it is therefore necessary after the (incomplete) integration to restore the cutoff parameter to its initial value by a renormalization of lengths; and also to readjust the value of the coefficient of the term in k^2, which will necessitate a renormalization of the order parameter.

This procedure yields a whole set of correspondences depending continuously on the value of s, this being a measure of the range over which the integration has been extended. Such a correspondence can be specified by a transformation between 'Hamiltonians'

$$\mathscr{H}' = f_s\{\mathscr{H}\} \tag{4.4}$$

which gives the parameters of $\mathscr{H}'$ in terms of those of $\mathscr{H}$. With each 'Hamiltonian' $\mathscr{H}$, i.e. with each system, we associate a point in a multidimensional parameter space. The set of all transforms $\mathscr{H}'$ of $\mathscr{H}$ is associated with a trajectory in this space. The product of two transformations (4.4) is well-defined through

$$f_{s_1 . s_2} = f_{s_1} . f_{s_2} \tag{4.5}$$

The 'group' of renormalizations consists of the set of all such transformations f_s along with their products; strictly speaking it is only a semigroup since the integrations involved are operations without an inverse.

The three operations of the renormalization group, and the dictionary of correspondences

We have now defined three operations:

1. Integration over part of the range, i.e. integration over the rapid, large wavenumber, fluctuations of the field variable†. This operation contains most of the physics (since an integration over the entire range would yield the full partition function); but it is easier said than done unless the problem is soluble by other means.

2. Readjustment of the cutoff parameter. This operation is motivated by the need to compare only like with like, which in real space means comparing problems defined on the same lattice; it amounts to a simple transformation of the scale of length, i.e. contraction of all lengths by a factor s.

3. Renormalization of the field variable so that the coefficient of the term in $k^2 M^2$ reverts to its initial value. The motivation for this step is less clear than for the first two; though the operation is obviously possible, there being no impediment to renormalizing the field variable, two questions arise nevertheless: is this step useful, and is it indispensable? The beginnings of an answer to the first question can be seen in the application to the Gaussian model in Section 4.3. The second question raises the more general problem of the uniqueness of the renormalization group, which is a difficult one and has barely been broached (see Section 5.6). Here we confine ourselves to the following remark: if the functional integral is changed into an integral over discrete lattice sites, then the coefficient of the term in $k^2 M^2$ determines the relative weighting of the fluctuations in the lattice integral. By keeping this coefficient constant one satisfies the physically reasonable requirement of maintaining constant weighting for equivalent sites in all systems.

In the light of these remarks we can begin to draw up a dictionary of correspondences between the systems defined by $\mathcal{H}$ and by $\mathcal{H}' = f_s\{\mathcal{H}\}$. Primed lengths are reduced with respect to the initial lengths by a factor s:

$$\mathbf{x} \to \mathbf{x}' = \mathbf{x}/s \tag{4.6}$$

In particular, we have for the correlation length

$$\xi \to \xi' = \xi/s \tag{4.7}$$

For the thermodynamic potential per unit volume, G, the correspondence is

$$G \to G' = s^d G$$

In fact the thermodynamic potential, which is directly related to Z, is left

---

† Translator's note. We shall refer to this operation as a partial integration: (it has nothing to do and should not be confused with 'integration by parts').

unchanged by the partial integration in the first step; the factor s^d derives from the second step which changes the unit of volume, d being the dimensionality of space. The contraction of lengths increases the density G'.

In general the rules of correspondence for the order parameter and its correlation functions are less simple: one can write

$$\mathbf{M(x)} \rightarrow \mathbf{M'(x')} = \lambda \cdot \mathbf{M(x)},$$

but as a rule λ depends simultaneously not only on s but also on all the parameters specifying $\mathscr{H}$. Locally, i.e. in the neighbourhood of a definite point in parameter space, one need retain only the dependence on s; then one can define the exponent d_ϕ, called the anomalous dimension of the order parameter, by

$$\lambda \sim s^{d_\phi}$$

The form of this power law in s follows from the relation $\lambda(s_1 \cdot s_2) = \lambda(s_1) \cdot \lambda(s_2)$, which in turn follows from (4.5). Thus one obtains for the order parameters the correspondence rule

$$\langle \mathbf{M} \rangle \rightarrow \langle \mathbf{M'} \rangle = s^{d_\phi} \langle \mathbf{M} \rangle$$

and for the correlation function $\Gamma(R) = \langle \mathbf{M(0)M(R)} \rangle$,

$$\Gamma(R) \rightarrow \Gamma' \left(R' = \frac{R}{s} \right) = s^{2d_\phi} \Gamma(R)$$

We must stress again that as a rule the value of the exponent d_ϕ is defined only 'locally', i.e. with respect to a particular region of parameter space.

4.2 Significance of the operations in real space

Before we begin to implement our programme it is useful to pause for a moment in order to clarify its significance.

The effect of the first step is to reduce the number of degrees of freedom by eliminating those included in the range of the partial integration. Since they are the ones with high wavenumbers, there is an obvious similarity with Kadanoff's approach mentioned in the introduction (Section 1.2). By working in reciprocal space rather than in real space like Kadanoff with his blocks of spins, we avoid some of the complications that beset him. Nevertheless the physics behind the integration over fluctuations having wavenumbers $1/sa < k < 1/a$ is the same as the physics behind the formation of blocks of spins having volume $(sa)^d$ in real space.

The effect of the second step is to change the lattice formed by the blocks back into the original lattice; this is a contraction by a factor s. Following as it does on the first step, the second step restores the spatial density of degrees of freedom to its initial value. But, had we started with a system of finite volume, then at this stage we would be left with a system whose volume is smaller. The number of degrees of freedom has indeed been reduced, and the irreversible nature of this elimination entails that the set of all such transformations constitutes not a group but a semigroup.

The significance of the third step is the same in real as in reciprocal space; we need only note that the renormalizations of $M(\mathbf{x})$ and $M(\mathbf{k})$ differ by a factor s^d because Fourier transformation involves an integration over space.

Throughout the arguments which follow it will be useful to keep in mind the cycle consisting of these three operations;

1. Integration (partial; formation of blocks);
2. Contraction (in real space, the blocks are rescaled to the original lattice);
3. Renormalization (of the magnitude of the spins).

They are illustrated in Figure 4.1.

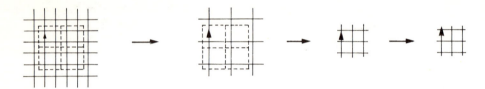

Figure 4.1. The three operations of the renormalization group in real space. Step 1: integration; step 2: contraction; step 3: renormalization of the magnitudes of the spins

4.3 The Gaussian model

The Gaussian model is obtained by setting $u_0 = 0$ in the expression (4.2) for the 'Hamiltonian' $\mathscr{H}$:

$$\mathscr{H} = \mu_0 + \tfrac{1}{2} \int_{|\mathbf{k}| < 1/a} (r_0 + k^2)|M(\mathbf{k})|^2 \, d\mathbf{k} \tag{4.8}$$

This model is directly soluble and the solution was given in Section 2.2; (notice some trivial notational differences, (factors of β), which arise because we are now using the 'Hamiltonian', instead of the free energy as in Section 2.2).

We shall now rederive the results of the Gaussian model by renormalization-group methods; while this is not a great achievement in itself, it will be most useful for clarifying what is actually involved in implementing the operations described above.

In the present case the first step is trivial, because the integrations over the $M(\mathbf{k})$ are independent of one another; then the definition (4.3) of $\mathscr{H}$ gives straightforwardly, up to a constant,

$$\mathscr{H}' = \tfrac{1}{2} \int_{|\mathbf{k}| < 1/sa} (r_0 + k^2)|M(\mathbf{k})|^2 \, d\mathbf{k} \tag{4.9}$$

The second step amounts to changing variables in the integral (4.9); letting $\mathbf{k} = \mathbf{k}'/s$, we find

$$\mathscr{H}' = \tfrac{1}{2} \int_{|\mathbf{k}'| < 1/a} (r_0 + k'^2 . s^{-2}) s^{-d} |M(\mathbf{k}')|^2 \, d\mathbf{k}' \qquad (4.10)$$

In the third step the coefficient of the term in $k^2 M^2$ is readjusted by setting $M = \zeta M'$, the factor ζ being chosen to satisfy

$$s^{-(d+2)} \zeta^2 = 1, \qquad \zeta = s^{(d+2)/2} \qquad (4.11)$$

Then the 'Hamiltonian' $\mathscr{H}'$ can be written in its final form

$$\mathscr{H}' = \mu_0' + \tfrac{1}{2} \int_{|\mathbf{k}'| < 1/a} (r_0' + k'^2) |M'(k')|^2 . d\mathbf{k}' \qquad (4.12)$$

where we have introduced $r_0' = r_0 . s^2$; this is to be compared to the initial 'Hamiltonian' $\mathscr{H}$ in (4.8). For the Gaussian model we can now complete the dictionary of correspondences from the preceding section:

$$\xi' = \frac{\xi}{s}, \quad G' = s^d . G$$

$$M'(\mathbf{x}') = s^{d\phi} . M(\mathbf{x})$$

$$s^{d\phi} = s^d . \zeta^{-1} \text{ (since } M(\mathbf{x}) \sim \int M(\mathbf{k}) \exp(i\mathbf{k} . \mathbf{x}) \, d\mathbf{k}), \quad \text{whence} \quad d_\phi = \frac{d-2}{2}$$

$$r_0' = s^2 . r_0, \qquad \text{i.e.} \qquad (\Delta T)' = s^2 . (\Delta T)$$

By eliminating s from $\xi' = \xi/s$ and $(\Delta T') = s^2(\Delta T)$, we then obtain

$$\xi \sim (\Delta T)^{-1/2}, \qquad \text{whence} \qquad v = \frac{1}{2}$$

More generally, from the relation

$$\Gamma'\left(\frac{R}{s}, r_0'\right) = s^{2d\phi} \Gamma(R, r_0)$$

plus the fact that (4.8) and (4.12) imply

$$\Gamma'(R', r_0') = \Gamma(R', R_0')$$

one obtains for the correlation function the homogeneity rule

$$\Gamma(R, \Delta T) = s^{-(d-2)} . \Gamma\left(\frac{R}{s}, s^2 \Delta T\right)$$

One can see without more ado that such arguments will yield homogeneous expressions for the thermodynamic potential as well as for the correlation function, with the critical exponents appropriate to the Gaussian model. We leave it as an exercise to generalize these results to the situation where there is an external field conjugate to the order parameter; and turn to some comments on what we have done.

The solvability of the Gaussian model stems from the fact that the first step becomes trivial; by contrast, in the general case it is precisely the first step that is most difficult. But here, because the first step is trivial, the calculation is dominated by the effect of the second step; and because this is a simple contraction, it is not surprising that the Gaussian results are just those of a naïve dimensional analysis (see Section 3.2). In the Gaussian case, the transformations of the renormalization group reduce to simple changes of scale.

Other simplifications also obtain; we see that the factor which renormalizes the magnitude of the spins is independent of the parameters in $\mathscr{H}$, and that the number of parameters needed to specify $\mathscr{H}'$ is no greater than it is for $\mathscr{H}$. Of course all these simplifications have a common source in the trivial nature of the first step. Notice that the third step has enabled $\mathscr{H}'$ to be written in a form that is as close as possible to $\mathscr{H}$, by compensating for the effects of the contraction on the term involving $k^2 M^2$; this illustrates very clearly the motivation for the third operation.

Next we turn to parameter space and to the trajectories in this space. Since there is only one parameter, r_0, parameter space reduces to a line (Figure 4.2), which is effectively the temperature axis, $(r_0 = T - T_0)$. As s increases, $r'_0 = s^2 \cdot r_0$ increases, so that the trajectories move away from the origin. The negative axis $r_0 < 0$ is an unphysical region, because the integrals are not defined there; (the Gaussian model not being defined for $T < T_0$). By contrast, the positive axis $r_0 > 0$ is a physical region. The origin is a fixed point, for if $r_0 = 0$, then $r'_0 = 0$ whatever the value of s. The 'Hamiltonian' $\mathscr{H}$ corresponding to this point remains invariant under the transformations of the renormalization group.

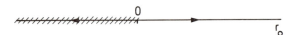

Figure 4.2. Parameter space for the Gaussian case. The cross-hatched semi-axis $r_0 < 0$ is an unphysical region. The Gaussian fixed point is at the origin ($r_0 = 0$)

After this detailed dimensional analysis of the solvable Gaussian model, we proceed to the main features of the general case.

4.4 Fixed points and their operator algebras

The Gaussian case dealt with in the last section corresponds to the limit of the expression (4.2) when $u_0 = 0$. When $u_0 \neq 0$, many simplicities of the Gaussian case disappear. In particular, the first step generates all manner of new coupling constants in $\mathscr{H}'$, because it is no longer possible to integrate over the fluctuations of every $\mathbf{M(k)}$ independently. Moreover, the renormalization of $\mathbf{M(k)}$ in the third step depends as a rule both on s and on the particular system being considered. But in spite of this complexity we shall show how one can generalize some of the concepts that have emerged from our study of the Gaussian case.

Consider parameter space; in general the parameters are r_0, u_0, plus many others which correspond to the couplings generated by the first step. The Gaussian case corresponds to the special subspace $u_0 = \cdots = 0$, r_0 arbitrary; this is a one-dimensional subspace (see Figure 4.2), and it contains a fixed point at $r_0 = 0$, called the Gaussian fixed point; this fixed point corresponds to a system having infinite correlation length, namely to the Gaussian model taken at its critical point.

In the (now enlarged) parameter space, it is conceivable that other fixed points may exist. We shall assume that some do exist, postponing the proof to Chapter 5, and examine the consequences.

An arbitrary point in parameter space is specified by its coordinates $\{\mu_i\}$; it is associated with a 'Hamiltonian' defined by

$$\overline{\mathscr{H}} = \sum_i \mu_i \cdot O_i$$

where the operator O_i is conjugate to the parameter (i.e. field) μ_i.

Let P* be a fixed point of the renormalization group, having coordinates $\{\mu_i^*\}$ and the associated 'Hamiltonian' $\overline{\mathscr{H}}^* = \sum_i \mu_i^* O_i$. By saying that P* is a fixed point, we mean

$$\overline{\mathscr{H}}^{*\prime} = f_s\{\overline{\mathscr{H}}^*\} = \overline{\mathscr{H}}^*$$

Applying our dictionary of correspondences to this special case we obtain

$$\xi' = \frac{\xi}{s} = \xi, \qquad \text{whence} \qquad \xi = \infty, \quad \text{(apart from the trivial case } \xi = 0\text{);}$$

$$\Gamma'\left(\frac{R}{s}\right) = s^{2d_\phi} \cdot \Gamma(R) = \Gamma\left(\frac{R}{s}\right), \qquad \text{whence} \qquad \Gamma(R) \sim \frac{1}{R^{2d_\phi}}$$

where the exponent d_ϕ has a specific value, namely its value at the point P*. Accordingly, to every fixed point (of the type $\xi = \infty$) there corresponds a critical system, lacking a characteristic scale of length, and having a correlation function which decreases like a power of distance.

To make further headway we now consider what happens at points P near P*, or in other words for 'Hamiltonians' $\overline{\mathscr{H}}$ differing little from $\overline{\mathscr{H}}^*$; thus we write

$$\overline{\mathscr{H}} = \overline{\mathscr{H}}^* + \delta\overline{\mathscr{H}} \qquad \mu_i = \mu_i^* + \delta\mu_i$$

Under a transformation of the renormalization group $\overline{\mathscr{H}}$ changes to $\overline{\mathscr{H}}'$, such that

$$\overline{\mathscr{H}}' = \overline{\mathscr{H}}^* + \delta\overline{\mathscr{H}}' \qquad \mu_i' = \mu_i^* + \delta\mu_i'$$

where the $\delta\mu_i'$ are obtainable from the $\delta\mu_i$ through relations expressible in the form

$$\delta\mu_i' = \sum_j A_{ij}\delta\mu_j$$

If the $\delta\mu_i$ and $\delta\mu_i'$ are small enough we can approximate by linearizing these relations; this amounts to replacing the coefficients A_{ij}, which in general are functions of the μ_i, by their values at the fixed point $\{\mu_i^*\}$. Then the coefficients A_{ij} define a matrix whose eigenvalues λ_i and eigenvectors g_i can be determined; they satisfy

$$g_i' = A \cdot g_i = \lambda g_i$$

The eigenvalues are functions of s; since the product of two transformations is defined (see 4.5), these functions must satisfy

$$\lambda_i(s_1 s_2) = \lambda_i(s_1) \cdot \lambda_i(s_2), \qquad \lambda_i(1) = 1$$

which in the neighbourhood of the fixed point imposes power-law behaviour on the $\lambda_i(s)$:

$$\lambda_i(s) = s^{y_i}$$

Around each fixed point we thus define an operator algebra: the operators in question are conjugates of the scaling fields, i.e. of those fields which, under the operations of the renormalization group, transform by simple multiplicative renormalization. In each algebra we can arrange the operators into a hierarchy according to decreasing values of the exponents y_i, which are the anomalous dimensions of their conjugate fields.

If $y_i > 0$, the field g_i increases as s increases; one says that the field g_i, and by extension its conjugate operator, are relevant.

If $y_i = 0$, then within the linearized approximation, the field g_i remains constant; we say that field and operator are marginal.

If $y_i < 0$, the field g_i decreases as s increases; we say that field and operator are irrelevant.

Here we have introduced the assumption that the algebra of the scaling operators O_i is closed, i.e. that any operator O can be expressed as a linear combination of the O_i. In particular, we adopt the hypothesis that any product of two operators O_i can be written as such a linear combination, and refer to this as the reduction hypothesis for operator products. Notice that there is no reason why the matrix A should be symmetric, whence the closed nature of the algebra is not guaranteed in advance. Nevertheless the hypothesis seems to apply in all cases that have been studied.

One begins to see at this stage how the theory meshes with the phenomenological homogeneity rules discussed in Chapter 3. It remains only to deduce the consequences of the dictionary of correspondences. We note the general rule

$$G' = s^d \cdot G$$

plus the fact that in the neighbourhood of the fixed point one has

$$G' = G(g_i'), \quad \text{where} \quad g_i' = s^{y_i} g_i$$

From these we obtain the homogeneity rule

$$G(s^{y_i} g_i) = s^d G(g_i)$$

which should be compared to Equations (3.36) in Chapter 3. For the correlation function we obtain similarly

$$\Gamma\left(\frac{R}{s}, s^{y_i} g_i\right) = s^{2d_\Phi^*} \Gamma(R, g_i)$$

also to be compared to (3.36).

4.5 The topology of trajectories near a fixed point: stability of fixed points

Consider a fixed point P* in parameter space. In the neighbourhood of P* we define local axes corresponding to the scaling fields discussed in the last section. Along axes which correspond to relevant fields the trajectories behave centrifugally, i.e. they move away from P*; along axes which correspond to irrelevant fields the trajectories behave centripetally. Behaviour along axes corresponding to marginal fields, if any, must be looked at more closely.

Catchment areas and the critical surface

We shall group the local axes into sets spanning subspaces of increasing dimensionality. Consider first the set of centripetal axes; locally they span a subspace of parameter space called the catchment area of P*, the locus of points P which converge towards P*.

The catchment area of P* is a subspace also of the critical 'surface'; the critical 'surface' is that subspace of parameter space which contains all points representing systems with infinite correlation lengths. It is obvious that every point belonging to the catchment area of P* also belongs to the critical surface, since the correlation length can only decrease under the transformations of the renormalization group. Accordingly, near P* the critical surface is spanned by the set of all centripetal axes plus, possibly, a certain number of centrifugal axes. In the directions of such centrifugal axes the trajectories move away from P* and eventually approach other fixed points also belonging to the critical surface.

Lastly there are centrifugal axes along directions in which the trajectories move away from the critical surface through points having decreasing correlation lengths.

Criticality conditions, and conditions for convergence to a fixed point

With every point P near a fixed point P* we associate a 'Hamiltonian' $\mathscr{H}$ written as

$$\mathscr{H} = \mathscr{H}^* + \mu_0 + \sum_i g_i O_i \tag{4.13}$$

The fields g_i are scaling fields that are renormalized by factors s^{y_i}, y_i being the anomalous dimension of the field g_i.

The condition that the system represented by $\mathscr{H}$ be critical, i.e. that P be on the critical surface, is that all those relevant fields should vanish which lead away from the critical surface. This imposes certain constraints, called criticality conditions.

Assuming that the criticality conditions are satisfied, certain other conditions must be satisfied in addition if $\mathscr{H}$ is to converge to $\mathscr{H}^*$, or in other words if the critical regime of $\mathscr{H}$ is to be governed by the fixed point P*. These additional conditions are that all those relevant fields, too, should vanish which lead away from P* while still remaining on the critical surface.

Before pursuing these conditions in detail, it is important to classify the operators O_i according to their symmetry properties.

Symmetries of operators

The important symmetries of operators are those under rotations in real space and in the space spanned by the order parameter. Such symmetry properties are conserved under the transformations of the renormalization group, as one can readily verify by analysing the three group operations described in Section 4.1.

The next task is to review the more important fields from the point of view of the symmetries of their conjugate operators. We begin with isotropic (i.e. scalar) operators. Amongst these there is one which plays an altogether special role, namely the unit operator, conjugate to the field μ_0. In Equation (4.13) we displayed it separately from the others, for even though the field μ_0, having anomalous dimension

$$y_0 = d,$$

is relevant, it contributes only an additive constant to the free energy, (recall our subdivision of the free energy G into a regular and a singular part).

The isotropic scaling operator next in relevance to the unit operator is denoted by O_1 or O_E. The second symbol alludes to the fact that O_E forms part of the energy operator when the latter is expressed as a scaling operator. The scaling field conjugate to O_E is proportional to $\Delta T = T - T_c$: thus one can say that ΔT has the anomalous dimension y_E of the field coupled to O_E. In practice one always has $y_E \geqslant 0$; the condition that the field ΔT vanish sets the temperature scale. It is a criticality condition well known from experiment; indeed, for an ordinary critical point it is the only criticality condition, provided the field conjugate to the order parameter vanishes. Hence, for a fixed point corresponding to an ordinary critical point, one can predict on empirical grounds that, apart from the unit operator, the energy is the only relevant scalar operator. This prediction is confirmed in all cases where calculations are possible.

There exist also fixed points having a second relevant scalar operator. To observe critical behaviour in such cases one must ensure the vanishing of two fields. This happens at tricritical points like the one shown in Figure 1.3; to reach this tricritical point one must choose appropriate values both for the temperature and for the uniform magnetic field (in this case the latter is not the field conjugate to the order parameter). The generalization to polycritical points is obvious.

Amongst the fields which break the symmetry there is a special one, namely the field conjugate to the order parameter, whose anomalous dimension

$$y_h = d - d_\phi$$

is always positive in practice; hence the vanishing of the field conjugate to the order parameter is one criticality condition that must always be imposed.

In Chapter 7, and later, we shall encounter several examples of symmetry-breaking operators relevant to the isotropic fixed point. The existence of such relevant fields underlies the classification of systems into simple and complex. In effect these relevant fields define centrifugal axes which generally lead to other fixed points, associated with different critical behaviours.

As regards the specification of scaling operators with various symmetries, there are certain simplifications for large n which allow one to progress quite far; we refer to Ma (1974) for a clearer picture of what is involved in an operator algebra with its hierarchy of anomalous dimensions. One should remember that such an algebra is defined locally, i.e. at some definite point of parameter space, and that each fixed point accordingly has its own algebra.

Stability of fixed points

The discussion above leads to the following picture. Consider systems with interactions that are isotropic in real space and also in the space spanned by the order parameter. In the parameter space associated with these systems we observe fixed points that can be classified as follows.

Singly unstable fixed points (i.e. unstable with respect to temperature), corresponding to ordinary critical points; doubly unstable fixed points (i.e. unstable with respect to temperature and to one other relevant field), corresponding to tricritical points; and so on.

If parameter space is enlarged by the introduction of new fields, especially fields breaking the symmetry, one can observe that the isotropic fixed points are unstable with respect to some of these fields. When speaking of the degree of stability of a fixed point, it is essential therefore to specify in advance the parameter space under consideration. The field conjugate to the order parameter provides a trivial but illuminating example. In the space spanned by the isotropic (scalar) parameters, an ordinary fixed point is singly unstable. But if the field conjugate to the order parameter is now included amongst the parameters, then an additional instability appears.

We have stressed that fields breaking the symmetry are possible causes of instability, but they are not the only such causes; others include, for instance, long-range forces (Chapter 10) and couplings to other degrees of freedom (Chapter 11). Summarizing, to solve a problem completely one must specify the effects of all possible fields, and must then determine all the fixed points and their stabilities. In order to do this one will proceed methodically, beginning with a study of the simplest possible topology in as restricted a parameter space as possible, and will then enlarge this space systematically.

4.6 Conclusion

We have seen how an approach through the renormalization group allows one, in principle, to understand universality, to prove the homogeneity rules discussed in Chapter 3, and to calculate the critical properties. But the renormalization group is not a panacea, and the next task is to understand the conditions under which the method can succeed. The next chapter shows why exact results for ordinary critical points are obtainable in the neighbourhood of the characteristic dimensionality $d = 4$.

Appendix 4.1

We apply the renormalization-group method to the one-dimensional Ising model.

Write the Hamiltonian for this model, with nearest-neighbour couplings, as

$$\mathscr{H} = -J \sum_i S_i S_{i+1}$$

where the 'spins' S_i can take the values ± 1; the index i indicates lattice sites. The free energy and correlation function for this model are very easily calculated. Before applying the ideas of the renormalization group we recall briefly the principles and results of the direct method.

One way to solve directly is to iterate; with Z_N the partition function of a chain of N spins, one notes that

$$Z_{N+1} = 2Z_N \cosh K, \qquad K \equiv J/T$$

which by iteration yields

$$Z_N = 2^{N+1}(\cosh K)^N$$

This leads to the free energy G per spin,

$$\tilde{G} = -\log(2\cosh K), \qquad \tilde{G} \equiv G/T$$

Another method uses the technique of transfer matrices; the partition function of a closed loop of N spins can be written as

$$Z_N = \text{Tr}(P^N)$$

where the transfer matrix P is given by

$$P = \begin{pmatrix} \exp(K) & \exp(-K) \\ \exp(-K) & \exp(K) \end{pmatrix}$$

In terms of the eigenvalues of P, $\lambda_1 = 2\cosh K$ and $\lambda_2 = 2\sinh K$ one obtains the free energy per spin

$$\tilde{G} = -\log \lambda_1 = -\log(2\cosh K)$$

and the correlation function

$$\Gamma(R) = (\tanh K)^R = \exp(-R/\xi)$$

where

$$\xi = \frac{1}{\log(\lambda_1/\lambda_2)} = -\frac{1}{\log(\tanh K)}$$

These results display some features specific to the model: no phase transition at finite temperature, an essential singularity (like $\exp(1/T)$) at zero temperature, etc. Because of these special features the application of the renormalization group is atypical in some respects; we shall comment on them at the end of this Appendix.

We start with the 'deimation' of the degrees of freedom by summing over fluctuations at, say, even-numbered sites, i.e. at one site out of two. Along the lines of Equation 4.3 we define

$$\exp(-\mathscr{H}') = \sum_{s_2,s_4,\ldots} \exp(-\mathscr{H})$$

But

$$\exp(-\mathscr{H}) = \langle S_1|P|S_2\rangle\langle S_2|P|S_3\rangle\langle S_3|P|S_4\rangle\langle S_4|P|S_5\rangle\ldots$$

whence

$$\exp(-\mathscr{H}') = \langle S_1|P^2|S_3\rangle\langle S_3|P^2|S_5\rangle\ldots \ .$$

In order to express $\mathscr{H}'$ in a form as like $\mathscr{H}$ as possible, we define

$$P(K') = \exp(-\mu_0)P^2(K)$$

which determines K' and μ_0 through

$$\exp(2K') = \cosh 2K$$

$$\mu_0 = K' + \log 2$$

For this dilatation by a factor 2 the rules of correspondence are

$$K \to K' = \tfrac{1}{2}\log\cosh(2K)$$

$$\tilde{G} \to \tilde{G}' = 2\tilde{G} + \mu_0$$

or in other words,

$$\tilde{G}(K') = 2\tilde{G}(K) + \mu_0(K)$$

Notice that the exact expression for $\tilde{G}$, derived at the beginning of this Appendix, certainly satisfies this recursion relation. The recursion relation for K shows that there are two fixed points, one at $K = 0$ (infinite temperature) which is a fixed point of the type $\xi = 0$, and the other at $K = \infty$ (zero temperature) which is of the type $\xi = \infty$.

It is easy to generalize to an arbitrary dilatation factor s, by noting that in terms of the eigenvalues of the transfer matrix the recursion relation becomes

$$\lambda(s) = \exp(-\mu_0(s))\lambda^s$$

Whence

$$2 \cosh K(s) = \exp[-\mu_0(s)](2 \cosh K)^s$$
$$2 \sinh K(s) = \exp[-\mu_0(s)](2 \sinh K)^s$$

this leads to

$$\tanh K(s) = (\tanh K)^s \qquad \exp[\mu_0(s)] = \frac{(2 \cosh K)^s}{2 \cosh K(s)}$$

One can now draw up the following dictionary of correspondences:

$$\xi \to \xi' = \xi/s$$
$$\tilde{G} \to \tilde{G}' = s\tilde{G}(K) + \mu_0(s)$$

and can write the recursion relation for K in differential form as

$$\frac{dK}{dl} = \left(\frac{\tanh K}{1 - \tanh^2 K}\right) \log \tanh K, \qquad s \equiv \exp(l)$$

In the respective neighbourhoods of the two fixed points ($K = 0$ and $K = \infty$), this differential equation takes the form

$$\text{small } K: \quad \frac{dK}{dl} \approx K \log K$$

$$\text{large } K: \quad \frac{dK}{dl} \approx -\frac{1}{2}$$

The last expression shows that $T = \dfrac{J}{K}$ is a marginal field, which entails $v = \infty$ and the existence of essential singularities.

After this brief survey we must comment on some special features. As in the Gaussian case (Section 4.3), the first step (the partial integration) does not introduce any new couplings; it is this which in both cases enables us to obtain exact results at all temperatures. On the other hand, and this is a simplification over and above the Gaussian case, the third step (renormalization of the spins) is now completely redundant, as a consequence of the fact that $\eta = 1$, whence $d_\phi = 0$. The price of these simplifications is the somewhat pathological nature of the results, which reduces the paedagogical value of the model. But even so, the one-dimensional Ising model, like the Gaussian model to which it is in some ways complementary, allows one to follow clearly and in detail some of the processes of the renormalization-group method.

General References

Fisher, M. E. (1974), *Rev. Mod. Phys.*, **46**, 587.
Ma, S. K. (1974), *Rev. Mod. Phys.*, **45**, 589.
Wegner, F. J. (1973), Summer School, Siikajärvi, Finland.

Wilson, K. G., Kogut, J. (1974), *Phys. Reports* 12 C, No 2.
Wilson, K. G. (1975), *Rev. Mod. Phys.*, **47,** 773.

Reference

Ma, S. K. (1974) *Phys. Rev. A*, **10,** 1818.

CHAPTER 5
In the neighbourhood of $d = 4$

5.1 On the characteristic dimensionality

In Chapter 2, and more particularly in Section 2.4, we introduced the concept of characteristic dimensionality, in the context of an argument about the validity of Landau theory. The Ginzburg criterion shows that there exists a characteristic dimensionality below which fluctuations are so important that, in the neighbourhood of the critical point, they destroy the self-consistency of classical theory.

In Section 2.5 we saw how the characteristic dimensionality can depend on the range of forces and on the kind of critical point in question. For an ordinary critical point and for short-range forces, the characteristic dimensionality is $d_c = 4$. The quickest way to this result is to substitute, in Josephson's law $vd = 2 - \alpha$, the classical values of the critical exponents, i.e. $v = \frac{1}{2}$ and $\alpha = 0$; this yields $d_c = 4$ as the dimensionality for which the classical exponents obey the scaling law.

In the present chapter we shall investigate the situation in the neighbourhood of the dimensionality $d = 4$, and shall be led to expansions of the critical exponents and of the scaling functions in powers of $\varepsilon = (4 - d)$. This strategy of approach through the neighbourhood of the characteristic dimensionality is not limited to the case $d_c = 4$. Similar expansions can be obtained around other values of d_c appropriate to cases with long-range forces or tricritical points. In this chapter we choose $d_c = 4$ purely to be definite, since this will serve perfectly well to illustrate what happens in general, besides being the most important case physically.

The Ginzburg criterion indicates how classical theory fails. In the topological language of renormalization-group theory, this failure reflects the instability of the Gaussian fixed point, resulting from an exchange of stabilities between this fixed point and another one, which we call the non-trivial fixed point. The topological situation to be described in this chapter is accordingly summed up as follows. For $d > 4$, the Gaussian fixed point is more stable than the non-trivial fixed point; as d tends to 4, the non-trivial fixed point moves towards the Gaussian, and coincides with the latter at $d = 4$. For $d < 4$ the non-trivial fixed point moves away from the Gaussian fixed point; but at coincidence their stabilities were exchanged, and now it is the non-trivial fixed point which is more stable. The non-classical values of the exponents are governed by this non-trivial fixed point. It is the proximity of the non-trivial to the Gaussian fixed point which makes it possible to calculate successfully near $d = 4$ and to obtain expansions in powers of $\varepsilon = 4 - d$.

5.2 Stability of the Gaussian fixed point

We start from Ginzburg–Landau theory; the standard form of the 'Hamiltonian' (see Section 4.1) is

$$\mathscr{H} = \mu_0 + \frac{1}{2} r_0 \sum_i M_i^2 + u_0 \left(\sum_i M_i^2 \right)^2 + \frac{1}{2} \sum_i (\nabla M_i)^2 \tag{5.1}$$

The Gaussian model is obtained by setting $u_0 = 0$, and the Gaussian fixed point is at $r_0 = u_0 = 0$. The topology of trajectories near this point is easily determined; indeed we saw in Section 3.2 that with respect to the Gaussian fixed point the dimensions of the fields assume their canonical values. Thus the field variable M renormalizes according to

$$M \sim s^{d_\phi} \quad \text{where} \quad d_\phi = \frac{d - 2}{2}$$

the operators M^p renormalize according to

$$M^p \sim s^{x_p} \quad \text{where} \quad x_p = p d_\phi \, (p \ge \varepsilon 1)$$

and the conjugate fields u_p, (contributing to the 'Hamiltonian' terms in $u_p M^p$), renormalize according to

$$u_p \sim s^{y_p} \quad \text{where} \quad y_p = d - x_p = d - p d_\phi$$

Note that in view of (5.1) we have by definition $r_0 = u_2$ and $u_0 = u_4$.

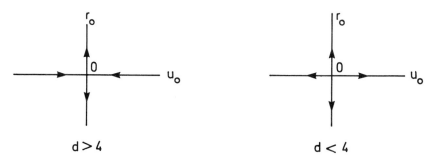

Figure 5.1.   Topology of trajectories near the Gaussian fixed point, for $d > 4$ and for $d < 4$

It follows in particular that

$$\Delta T \sim s^{y_2} \quad \text{where} \quad y_2 = 2, \quad \text{whence} \quad v = \tfrac{1}{2},$$

$$u_0 \sim s^{y_4} \quad \text{where} \quad y_4 = 4 - d.$$

We see that for $d > 4$ the field u_0 is irrelevant ($y_4 < 0$); for $d = 4$ it is marginal ($y_4 = 0$); and for $d < 4$ it is relevant ($y_4 > 0$). Figure 5.1 summarizes the topology of the trajectories in the projection of parameter space onto the (r_0, u_0) plane.

Dimensionality 4 thus marks the change in stability of the Gaussian fixed point with respect to the field u_0. The generalization to other fields is obvious; the Gaussian fixed point changes its stability with respect to a field u_p when $y_p = d - p\left(\dfrac{d-2}{2}\right) = 0$, or in other words when $d = \dfrac{2p}{p-2}$. We recognize the expression from Section 2.5 giving the characteristic dimensionality for arbitrary p.

5.3 The differential equations of the renormalization group

When we perform the transformations of the renormalization group as defined in Chapter 4, we find that the parameters r_0 and u_0 renormalize subject to the differential equation

$$\frac{dr_0}{dl} = 2r_0 + bu_0(1 - r_0) \tag{5.2}$$

$$\frac{du_0}{dl} = \varepsilon u_0 - cu_0^2 \tag{5.3}$$

Here we have defined $s = e^l$, $\varepsilon = 4 - d$, $b = 16(n + 2)$, $c = 16(n + 8)$; n is the dimensionality of the order parameter. These differential equations are approximations appropriate when ε, r_0 and u_0 are all small, since in writing each equation we have dropped terms of higher order in these variables. The renormalization procedure generates further coupling constants in addition to r_0 and u_0, which have likewise been neglected because they are small under the stated conditions; thus the discussion can be restricted to the subspace of parameter space spanned by r_0 and u_0. We shall start by examining the consequences of these results, postponing their derivation to Section 5.6.

Gaussian and non-trivial fixed points

One searches for fixed points by setting $\dfrac{dr_0}{dl} = 0 = \dfrac{du_0}{dl}$ in equations (5.2) and (5.3). This yields two solutions. The first, ($r_0^* = u_0^* = 0$) corresponds to the Gaussian fixed point with which we are already familiar; the second solution $\left(r_0^* = -\dfrac{b}{2c}\varepsilon, u_0^* = \dfrac{\varepsilon}{c}\right)$ corresponds to a new, the non-trivial, fixed point. In actual fact the non-trivial fixed point is not located in the plane (r_0, u_0); it has other coordinates that are non-zero, though they are negligible when ε is small, thus allowing us to consider only its projection on the plane (r_0, u_0).

In the neighbourhood of each fixed point one sets $r_0 = r_0^* + \delta r_0$, and $u_0 = u_0^* + \delta u_0$; by keeping only terms linear in δr_0 and δu_0 one obtains a set of two simultaneous linear differential equations which can be written as

$$\begin{pmatrix} \dfrac{d\delta r_0}{dl} \\[2mm] \dfrac{d\delta u_0}{dl} \end{pmatrix} = A \begin{pmatrix} \delta r_0 \\ \delta u_0 \end{pmatrix}$$

The matrix A_g appropriate to the Gaussian fixed point is

$$A_g = \begin{pmatrix} 2, b \\ 0, \varepsilon \end{pmatrix}$$

Its eigenvalues are $y_1 = 2$ and $y_{II} = \varepsilon$, in agreement with the results obtained by the direct method in Section 5.2. The matrix A_{nt} appropriate to the non-trivial fixed point is

$$A_{nt} = \begin{pmatrix} \left(2 - \dfrac{b}{c}\varepsilon\right), & b\left(1 + \dfrac{b}{2c}\varepsilon\right) \\[2mm] 0, & -\varepsilon \end{pmatrix}$$

with eigenvalues $y_1 = 2 - \dfrac{b\varepsilon}{c}$ and $y_{II} = -\varepsilon$.

The larger of the two eigenvalues, denoted by y_1 (or by y_E in the notation of Section 4.5), is positive. The exchange of stabilities, illustrated in Figure 5.2, is clearly reflected in the dependence of the sign of the second eigenvalue y_{II} on the choice between the two fixed points and on the sign of ε. Figure 5.2 complements Figure 5.1, and is basic to all that follows.

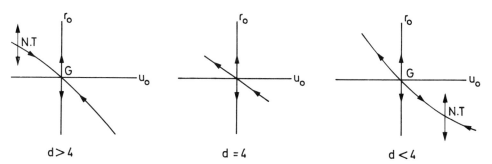

$$d>4 \qquad\qquad d=4 \qquad\qquad d<4$$

Figure 5.2. The exchange of stabilities between the Gaussian (G) and the non-trivial (NT) fixed points. For $d > 4$, the Gaussian is more stable than the non-trivial fixed point. For $d = 4$ the two points coincide. For $d < 4$ the stabilities are interchanged

Parameter space

In the plane (r_0, u_0), there is an important line, namely the critical line, which is the locus of points representing systems with infinite correlation lengths. Insofar as we can consider the non-trivial fixed point as located in the plane (r_0, u_0), this line passes through both fixed points, and at each is tangent to the slow axis through that point. By slow axis we mean the following: at each fixed point we can label the local axes by the corresponding eigenvalues of the matrix A; then the fast axis is said to be the one corresponding to the greater eigenvalue y_I, while the slow axis corresponds to the smaller eigenvalue y_{II}.

The scaling field corresponding to a given eigenvalue is a linear combination of the fields r_0 and u_0, and may be written as

$$g = (r_0 - r_0^*) + w(u_0 - u_0^*) \tag{5.4}$$

w is determined by comparison with the equation

$$\frac{dg}{dl} = y \cdot g$$

The equations of the tangents to the critical line are obtained from (5.4) by substituting the appropriate value of w and setting $g = 0$.

Knowledge of the critical line allows one to estimate the critical temperature T_c. Note that a small positive u_0 will lower T_c relative to its mean-field (Gaussian) value; the decrease is linear, hence analytic, in u_0:

$$\Delta T_c \sim -\frac{b}{2}u_0 \tag{5.5}$$

To visualize the system of trajectories it helps to consider an analogy with a contour map. The critical line can be visualized as a ridge; any deviation from it impels one further away in the same general direction. For $d > 4$, the non-trivial fixed point is a peak on this ridge, while the Gaussian fixed point is a saddle (mountain pass); for $d < 4$ these roles are exchanged; for $d = 4$ the fixed point, at the origin, is a point of inflexion. This geographical analogy is illustrated in Figure 5.3.

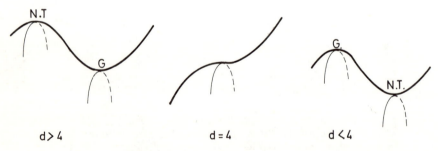

Figure 5.3. The exchange of stabilities, as in Figure 5.2, shown in relief

Homogeneity rules

Consider a system of dimensionality $d \leqslant 4$. Such a system is characterized by a physical line in the plane (r_0, u_0); as the temperature varies the point representing the system moves along this line, shown broken on Figure 5.4. The point of intersection P_c between the physical line and the critical line corresponds to the critical point of the system; P_c converges to the non-trivial fixed point which therefore governs the critical behaviour of the system.

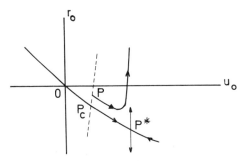

Figure 5.4.   The case $d \lesssim 4$. The broken line is the physical line of a system. The trajectory issuing from P_c converges to the fixed point P^*. The trajectory issuing from the point P, lying close to P_c on the physical line, first approaches P^* and then veers away

With respect to the local axes of the non-trivial fixed point, a point P near P_c is specified by two coordinates corresponding to the two scaling fields g_I and g_{II}, which have dimensions y_I and y_{II} respectively. Then the dictionary of the renormalization group gives the free energy, for instance, as

$$G(g_I, g_{II}) = s^{-d}G(g_I s^{y_I}, g_{II} s^{y_{II}})$$

In this case we have

$$g_I \sim \Delta T, \qquad y_I = 2 - \left(\frac{n+2}{n+8}\right)\varepsilon$$

$$g_{II} \sim (u_0 - u_0^*), \qquad y_{II} = -\varepsilon$$

so that we can write

$$G \sim (\Delta T)^{d/y_I} \cdot f\left(\frac{u_0 - u_0^*}{(\Delta T)^\phi}\right) \tag{5.6}$$

where

$$\phi = v y_{II} = -\frac{\varepsilon}{2}$$

Equation (5.6) gives the singular term, proportional to $(\Delta T)^{d/y_I}$, which dominates the free energy. Insofar as one can expand the function $f(x)$ in powers of

x for small x, we have

$$f(x) = f(0) + f'(0)x + \cdots$$

which shows that from (5.6) one can calculate also the corrections due to the irrelevant field $g_{\mathrm{II}} \sim (u_0 - u_0^*)$. The exponent ϕ is the crossover exponent associated with this field.

Other things being equal, it is clear that the further P_c is from the fixed point P^*, the greater the importance of the correction terms, and the narrower the temperature range of the critical region around T_c. In other words, when P_c is far from P^*, the only trajectories coming close to P^* are those which have issued from points very close to the critical line. Since $\Delta T = T - T_c$ is proportional to the distance between P and P_c this implies that behaviour characteristic of P^* is observed only over a narrow range of temperature.

We mentioned in Section 2.4 that for forces of long but finite range, the longer the range the smaller r_0 and u_0, and the narrower the critical region. To this we can now add that the longer the range, the further is the point P_c from P^*, which is consistent with the foregoing discussion.

Critical exponents

When $d = 4 - \varepsilon$, the non-trivial fixed point is the more stable. We have found the values of the two anomalous dimensions y_{I} and y_{II} corresponding to the two scaling fields defined in the neighbourhood of this fixed point. Recall that the dimension y_{I} determines the exponent v in view of the fact that

$$\xi \text{ renormalizes like } 1/s$$

$$\Delta T \text{ renormalizes like } s^{y_{\mathrm{I}}}$$

whence $\xi \sim \left(\dfrac{1}{\Delta T}\right)^{1/y_{\mathrm{I}}}$; but by definition $v = 1/y_{\mathrm{I}}$, so that

$$v = \frac{1}{2} + \frac{\varepsilon}{4}\left(\frac{n+2}{n+8}\right) \tag{5.7}$$

which is a formula for the first two terms of an expansion in power of $\varepsilon = 4 - d$. The dimension y_{II} has no bearing on any of the main critical exponents; as we have seen, it affects only the correction terms.

There exists one exponent which has not yet been mentioned in this chapter, but which is determined in the course of setting up the differential renormalization equations. This is the exponent d_ϕ which enters the renormalization of the order parameter. To first order in ε one finds

$$d_\phi = 1 - \frac{\varepsilon}{2} \quad \text{whence} \quad \eta = 0$$

The values of η and v, together with the scaling laws which follow from the homogeneity rules, lead to expressions for the other four main critical exponents, to first order in ε but for arbitrary values of n:

$$\gamma = 1 + \frac{\varepsilon}{2}\left(\frac{n + 2}{n + 8}\right)$$

$$\alpha = \frac{\varepsilon}{2}\left(\frac{4 - n}{n + 8}\right)$$

$$\beta = \frac{1}{2} - \frac{3}{2}\cdot\frac{\varepsilon}{n + 8}$$

$$\delta = 3 + \varepsilon$$

The special cases $n = -2$ and $n = \infty$

At this stage it is instructive to consider the special cases $n = -2$ and $n = \infty$, in order to compare the results just obtained to first order in ε with the exact results for arbitrary dimensionality d, which were derived by the direct method in Section 2.6.

For $n = -2$ we now find that the non-trivial fixed point has the coordinate $r_0^* = 0$; then Equation (5.5) implies that there is no shift in T_c. This is actually a general result holding for any value of d. As regards the main critical exponents, it is amusing to notice that they assume the same set of values at the two fixed points, Gaussian and non-trivial, even though these are quite distinct points of parameter space. Further, in view of the results obtained by the direct method it follows by continuity that this situation will persist for all values of the dimensionality d. Of course the two fixed points are not completely equivalent since their stabilities differ; nor is there identity between their respective scaling operators and their spectra of anomalous dimensions. (For an illustration of this fact see Section 7.2.)

For $n = \infty$, one sees here that the coordinate u_0^* of the non-trivial fixed point is of order $1/n$. This is related to the choice of a u_0 of order $1/n$ in the argument of Section 2.6. It is also easy to check that the critical exponents calculated here to order ε agree to this order with the exact values obtained directly.

5.4 The special case $d = 4$

When d is exactly equal to 4, the topology as shown in Figures 5.2 and 5.3 has the peculiarity that the two fixed points coalesce. From the point of view of the algebra of operators defined around the one fixed point which results, this coincidence is reflected by the existence of marginal operators. There is an evident connection between the two features, coalescence of two fixed points and existence of a marginal operator.

The marginal field here is u_0. For $d = 4$ the differential renormalization

Equations (5.2) and (5.3) reduce to

$$\frac{dr_0}{dl} = 2r_0 + bu_0(1 - r_0)$$

$$\frac{du_2}{dl} = -cu_0^2$$

The second equation integrates to

$$u_0(l) = \frac{1}{c(l + l_0)}, \qquad u_0(0) \equiv \frac{1}{cl_0}$$

Since $l = \log s$, this shows that instead of the usual power-laws like s^y we now have a renormalization proportional to logs.

The renormalization of the other scaling field $g = r_0 + \frac{b}{2}u_0$ is governed by the differential equation

$$\frac{dg}{dl} = (2 - bu_0)g = \left(2 - \frac{b}{c} \cdot \frac{1}{l + l_0}\right)g$$

This integrates to

$$g(l) = \exp(2l)\left(\frac{l}{l_0} + 1\right)^{-q} \cdot g(0)$$

where

$$q = \frac{b}{c} = \left(\frac{n + 2}{n + 8}\right)$$

The scaling field g is a measure of deviation from the critical line, and is proportional to ΔT. The dictionary of correspondences thus leads one to formulae like the following for the correlation length:

$$\xi(\Delta T) = \exp(l) \cdot \xi\left[\exp(2l)\left(\frac{l}{l_0} + 1\right)^{-q}\Delta T\right]$$

By choosing a value of l such that

$$\exp(2l)\left(\frac{l}{l_0} + 1\right)^{-q}\Delta T \sim 1$$

one finds the following:

$$\xi(\Delta T) \sim (\Delta T)^{-1/2}\left(1 - \frac{1}{2l_0}\log \Delta T\right)^{q/2}$$

This is an instructive expression. If initially u_0 is zero, i.e. $\frac{1}{l_0} = 0$, then it reduces to the Gaussian result as it must. For non-zero u_0 the dominant term is

$$\xi \sim (\Delta T)^{-1/2} (\log \Delta T)^{(n+2)/2(n+8)}$$

In other words the Gaussian power-law is multiplied by a logarithm raised to a fractional power. The crossover temperature between the mean-field (Gaussian) regime and this logarithmically corrected regime is obtained by setting

$$1 \sim -\frac{1}{2l_0} \log \Delta T$$

$$\Delta T \sim \exp\left(-\frac{1}{8(n+8)u_0}\right) \qquad (5.8)$$

Note that (5.8) is the special form for $d = 4$ of the Ginzburg criterion, which previously for $d < 4$ was written as

$$\Delta T \sim u_0^{2/\varepsilon}$$

and also that in (5.8), $u_0 \to 0$ entails $\Delta T \to 0$ as it must. This exponential form is typical of marginal cases; it recurs for instance in the expression for the Kondo temperature, which is the characteristic temperature in the problem of dilute magnetic alloys (Chapter 13).

From the expression for the correlation length one obtains the behaviour of the susceptibility χ; since $\chi \sim \xi^2$, one has

$$\chi \sim (\Delta T)^{-1} (\log \Delta T)^{(n+2)/(n+8)}$$

In order to find the logarithmic corrections to the other critical variables the only further information we need is how to renormalize the free-energy density G. For this we must take into account the renormalization of the term μ_0 in the 'Hamiltonian'. In principle one has to appeal to an additional differential equation, but in practice the result is very simple and can be obtained by considering the Gaussian special case:

$$G(l) \sim \exp(4l) \cdot l^{-1}$$

The factor l^{-1} in this renormalization of G is a source of further logarithmic corrections which are present even when u_0 vanishes. From G one finds the specific heat, which contains a contribution like

$$C \sim (\log \Delta T)^{(4-n)/(n+8)}$$

This diverges when $n < 4$.

Once we know how to renormalize the susceptibility χ and the free energy G, we can deduce how to renormalize the order parameter M and its conjugate field H, from $\chi = M/H$ and $G \sim MH$. This entails

$$M(l) \sim \exp(l) \cdot l^{-1/2}, \qquad H(l) \sim \exp(3l) \cdot l^{-1/2}$$

These expressions determine the behaviour of the order parameter below T_c

$$M(\Delta T) \sim |\Delta T|^{1/2} (\log |\Delta T|)^{3/(n+8)}$$

and also M as a function of H and T_c:

$$M(H) \sim H^{1/3} (\log H)^{1/3}$$

The examples above should suffice to show how the exponents of the logarithmic corrections are connected by relations that could be codified into scaling laws. One can check as an exercise that for $n = -2$ and $n = \infty$ the expressions obtained above do indeed reproduce the exact results obtainable directly by the methods of Section 2.6.

5.5 Domains in parameter space

Some values of parameters are unphysical. Consider the 'Hamiltonian' of Equation (5.1); we have already noted that when $u_0 = 0$, i.e. in the Gaussian case, the values $r_0 < 0$ are unphysical because $\mathscr{H}$ then has no lower bound, so that the functional integral which gives the partition function becomes meaningless. In the same way, for $u_0 < 0$ and r_0 arbitrary, the system has no physical significance unless we add a term $u_6 M^6$ with $u_6 > 0$. Accordingly there are unphysical domains in parameter space. For a more general understanding of the domain-structure of parameter space, it proves useful to widen the discussion of Section 5.3 by including the effects of the parameter u_6. The present section aims to do this.

Landau theory

To make the discussion more transparent we adopt the notation u_4 for the quantity conjugate to the order parameter and previously denoted by u_0. Consider the plane (u_4, u_6); Landau theory divides this plane into the three domains shown in Figure 5.5. The half-plane $u_6 < 0$ is unphysical; the quadrant $u_4 < 0$, $u_6 > 0$

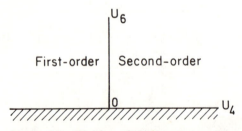

Figure 5.5. Division of the parameter space (u_4, u_6) into domains according to Landau theory. The cross-hatched region is unphysical; the top right-hand quadrant corresponds to second-order transitions, and the top left-hand quadrant to first-order transitions

corresponds to second-order transitions with classical exponents; the quadrant $u_4 < 0, u_6 > 0$ corresponds to first-order transitions. The semi-axis $u_4 = 0, u_6 > 0$ corresponds to tricritical points, with classical tricritical exponents (see Section 2.5).

Now imagine an r_0-axis perpendicular to the plane (u_4, u_6). In the parameter space (r_0, u_4, u_6) a physical system is represented by a line, namely its physical line; this line intersects the plane (u_4, u_6) at a point. If this point belongs to the quadrant $u_4 > 0$, $u_6 > 0$, then it corresponds to a critical point of the system, which undergoes a second-order transition. By contrast, if the point belongs to the quadrant $u_4 < 0, u_6 > 0$, then the situation is completely different; in that case the system undergoes a first-order transition which takes place for a positive value of r_0, or in other words out of the plane (u_4, u_6). Let the r_0-axis be directed so that high temperatures lie above the plane (u_4, u_6). As the temperature drops, the point representing the system descends along the physical line, and the transition occurs before the point reaches the plane (u_4, u_6).

The arguments above are appropriate in Landau theory. We shall now see the changes brought about by fluctuations.

The critical surface

Fluctuations lead to a shift in T_c, causing the critical surface to curve. Moreover, in order to represent the trajectories generated by the renormalization group, it would be necessary in principle to keep a fair number of parameters. But for display purposes we shall use diagrams drawn in the (u_4, u_6) plane to describe the topology of trajectories on the critical surface; one should keep in mind that the significance of such diagrams lies only in their topology.

The first task is to locate the non-trivial fixed point; expanding in powers of $\varepsilon = 4 - d$ one finds

$$u_4^* \sim \varepsilon, \qquad u_6^* \sim \varepsilon^2, \text{ etc.}$$

Insofar as one neglects components of higher order in ε, the non-trivial fixed point can now be inscribed on the (u_4, u_6) diagrams. In Figure 5.6 we sketch the situation for $d > 4$ and for $d < 4$. The broken lines are ridges separating two domains, one where trajectories converge on the Gaussian fixed point if $d > 4$ and on the non-trivial fixed point if $d < 4$, and the other where the trajectories apparently go off to infinity. The first domain corresponds to systems which undergo a second-order transition with critical behaviour governed by the attractive fixed point; the second domain, by continuity from Landau theory, is interpreted to correspond to systems undergoing a first-order transition which occurs before the critical surface is reached, i.e. before the correlation length has diverged. On the dividing line, trajectories converge on the non-trivial fixed point if $d > 4$ and on the Gaussian fixed point if $d < 4$.

Let us consider the diagram for $d < 4$ in more detail. Near the non-trivial fixed point we define two local axes, recognizing one as the slow axis, and the other as the fast axis, marked as such by a double arrow. If the physical line of a system

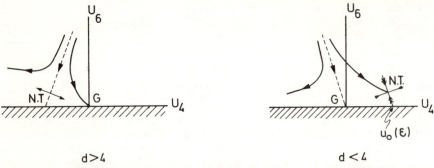

Figure 5.6. Topology of trajectories on the critical surface, parametrized by u_4 and u_6. The broken lines separate two domains, one where trajectories converge to a fixed point (second-order transitions), and one where trajectories travel out to infinity (first-order transitions). When $d < 4$, there exists a point on the u_4 axis with coordinate $u_0(\varepsilon)$, such that convergence to the non-trivial fixed point is fastest from $u_0(\varepsilon)$

intersects the critical surface at a point on the fast axis, i.e. at a point lacking a slow component, then the critical region will be wide. In particular, if we start with a 'Hamiltonian' having $u_6 = 0$, then there exists a special value $u_0(\varepsilon)$ of u_4 for which the critical region is exceptionally wide. The figure makes it clear that $u_0(\varepsilon)$ is indeed a function of ε, and that it differs from u_4^* to order ε^2. The next section shows what use can be made of $u_0(\varepsilon)$.

We should stress here that this entire argument has been constructed to suit values of d close to 4, which enables one to work in a very restricted subspace of parameter space. Nevertheless, by appeal to topological continuity one can visualize the changes that occur as d moves away from 4.

Thus, one can predict what happens when d decreases through $d = 3$. At that stage a new fixed point appears; it approaches the Gaussian fixed point from the region of negative u_6, coincides with it when $d = 3$, and for $d < 3$ moves away from it into the region of positive u_6. Once again there is an exchange of stabilities, connected in this case with the circumstance that u_6 is Gaussian-marginal when $d = 3$. When $d = 3 - \varepsilon$, the whole argument can be repeated, yielding, in particular, expansions of the tricritical exponents in powers of $\varepsilon = 3 - d$.

5.6 Calculational techniques

At this stage one should attempt actually to derive the renormalization equations on which all of the preceding arguments are based. We shall give a quick review of the manifold calculations that have been done and of their variants (e.g. recursion relations vs. differential equations, or sharp vs. smooth cutoff). Such variants raise the very important question whether the solutions are unique. This part of the discussion takes place in the framework of the second version of the renormalization group.

Soon after formulating the second version, Wilson (1972) proposed a different method, leading to expansions in powers of $\varepsilon = 4 - d$. This is a perturbation calculus implemented for a special value $u_0(\varepsilon)$ of the coupling constant u_0; though it lies beyond the confines of the renormalization group it is thanks to this method that we can understand why the renormalization-group procedure succeeds.

There exists furthermore yet another, the so-called parquet method, which sums certain classes of perturbation diagrams, and allows one to calculate the logarithmic corrections for $d = 4$ as well as the first term of the expansion in powers of $\varepsilon = 4 - d$.

Finally we shall say a few words about calculations performed in the framework of the first version of the renormalization group, and about the conclusions to which they lead.

The second version of the renormalization equations

The Equations (5.2) and (5.3) given above are not the form in which the renormalization equations were originally derived. Wilson (1971) in his first calculation, working with an *a priori* arbitrary dimensionality d, obtained approximate recursion relations for a finite value $\bar{s} = 2$ of s. Thus originally his equations were not written in differential form. Afterwards, Wilson and Fisher (1972) realized that these approximate recursion relations could be made rigorous in the neighbourhood of dimensionality 4; they yield expansions for the critical exponents which are exact to first order in $\varepsilon = 4 - d$, but not beyond.

However, the recursion relations were not very satisfactory to handle. For instance, because of the particular method used, one found that the position of the non-trivial fixed point depended on the value chosen for the discontinuity $\bar{s}$; though the method was saved by the lucky circumstance that the critical exponents were independent of $\bar{s}$, as were, more generally, the anomalous dimensions in the algebra of operators associated with the fixed point. Nevertheless a differential form seemed preferable even if harder to define. The difficulties are discussed in the review article by Wilson and Kogut (1974); they depend on the kind of cutoff used, sharp or smooth. Without going into details we note only that the difficulties have been surmounted; the differential Equations (5.2) and (5.3) have been derived by Wegner and Houghton (1973) using a sharp cutoff.

The differential equations given above are correct to order ε; the terms of order ε^2 can be found but they quickly become very complicated. In fact, for calculations to higher orders it is better to use other methods, like the expansion in powers of $u_0(\varepsilon)$, or the Callan–Symanzik equations mentioned below.

It remains to discuss the uniqueness of the results obtained by the different variants of the second version of the renormalization group. One observes that all variants do yield identical physical results, but one observes also that they yield different positions for the fixed points. More especially, in those methods which lead to differential equations the position of the non-trivial fixed point still depends on the kind of cutoff function employed. On physical grounds it is obvious that the critical exponents must not depend on this, and it is satisfactory to observe that they do not. Nevertheless, we still lack a systematic classification of all possible renormalization procedures (with their networks of trajectories in parameter space), and we have not yet understood the uniqueness of results obtainable by such procedures. For further discussion we refer to Jona-Lasinio (1973) and Wegner (1974).

Expansions in powers of $u_0(\varepsilon)$

This method starts with the expansion of the correlation function $\Gamma(k, r_0, u_0)$ in a double series in powers of u_0 and ε. One keeps track order by order of the shift in T_c due to u_0, so that Γ can be evaluated at the critical point in each order. Next one chooses the value of u_0, as a function of ε, so that the perturbation expansion can be re-summed into power-laws. We can construct an example by considering a series like

$$1 + a_1(u_0, \varepsilon)(\log x) + a_2(u_0, \varepsilon)(\log x)^2 + \cdots$$

Then we choose that value of u_0, if such a value exists, for which $a_2 = a_1^2/2$, enabling the expansion to be re-summed into $\exp(a_1 \log x) = (x)^{a_1}$. The condition $a_2 = a_1^2/2$ determines $u_0(\varepsilon)$, and the final value of the exponent is $a_1(\varepsilon) = a_1(u_0(\varepsilon), \varepsilon)$.

It is not at all clear *a priori* whether there exists such a value $u_0(\varepsilon)$ allowing the logarithmic terms of all orders to be re-summed into power laws. Wilson (1972) in his calculation verifies its existence successively to each order in ε:

$$u_0(\varepsilon) \sim \frac{\varepsilon}{n + 8} + O(\varepsilon^2)$$

For greater insight into $u_0(\varepsilon)$ it is useful to return to the discussion of Figure 5.6 in Section 5.5. We saw that there exists a special value $u_0(\varepsilon)$ of u_0, for which the critical region is exceptionally wide because the slow component vanishes. What Wilson's calculation shows is that for this value of u_0 the critical region is so wide that homogeneity rules apply at all temperatures, and are reflected even in the perturbation series in powers of u_0, in spite of the fact that this is a high-temperature expansion.

Arguments based on the renormalization group thus enable one to understand the existence of a special value $u_0(\varepsilon)$ for which the critical region is exceptionally large. Admittedly, from here it is still quite a step to the statement that for this value power laws are observable even in first-order perturbation theory. To convince himself that nevertheless this is so, we suggest that the reader consider the case with n large, which is exactly soluble, and where one can indeed confirm the existence of such a $u_0(\varepsilon)$.

The parquet method

The parquet method is a technique in perturbation theory for a partial resummation of diagrams; the term 'parquet' refers to the general appearance of the diagrams to be summed. The method, already mentioned in Section 1.2, is convenient for problems where logarithmic divergences appear in perturbation theory. From the viewpoint of the renormalization group it is suited to cases with a marginal operator. This is how Larkin and Khmelnitskii (1969) could derive the results of Section 5.4 by the parquet method well in advance of recent developments. Indeed, since then it has been shown that the same method will also yield the leading term in the ε-expansions of the critical exponents.

The most accessible introduction to the parquet method is that given by Roulet, Gavoret and Nozières (1969), in the context of the problem of threshold anomalies in X-ray absorption and emission by metals. In addition to the applications already mentioned in Section 1.2 (field theory and Kondo effect), we list the problem of one-dimensional metals and the various cases of marginal behaviour in critical phenomena, like tricritical points precisely at their characteristic dimensionality. We shall not pursue the parquet method any further, and comment only that it always seems worth translating its results into the language of the renormalization group.

First version of the renormalization group, and the Callan–Symanzik equations

We shall attempt an outrageously sketchy outline aimed at facilitating certain comparisons between the first and second versions of the renormalization group. Consider a 'Hamiltonian' like that in (5.1),

$$\bar{\mathscr{H}} = \frac{1}{2}r_0\mathbf{M}^2 + \frac{u_0}{4!}(\mathbf{M}^2)^2 + \frac{1}{2}(\nabla\mathbf{M})^2 \tag{5.9}$$

with a cutoff parameter $\Lambda = 1/a$ at high momenta (see Chapter 4). We carry out a number of formal manipulations on $\bar{\mathscr{H}}$, whose significance will appear later. Setting $\mathbf{M} = Z_3^{1/2}\mathbf{M}'$, (the so called wave function renormalization), we obtain

$$\bar{\mathscr{H}} = \frac{1}{2}r\mathbf{M}'^2 + \frac{ur^{(4-d)/2}Z_1}{4!}(\mathbf{M}'^2)^2 + \frac{1}{2}Z_3(\nabla\mathbf{M}')^2 + \frac{1}{2}(Z_3r_0 - r)\mathbf{M}'^2 \tag{5.10}$$

where we have defined

$$u_0 = r^{(4-d)/2} \cdot u \cdot Z \; Z_3^{-2} \tag{5.11}$$

The coefficient r is introduced into (5.10) in order to take account of the shift in T_c; it enters (5.11) for dimensional reasons, Z_1 and u thus becoming dimensionless.

In field theory one would talk about a Lagrangian $\mathscr{L}$, a bare mass m_0 and a physical mass m, the correspondence being given by

$$\bar{\mathscr{H}} \to -\mathscr{L}, \qquad r_0 \to m_0^2, \qquad r \to m^2$$

Starting from (5.10) one calculates the correlation function $\Gamma'(k)$ of the field variable $\mathbf{M}'$, and subjects it to the three renormalization conditions which define Z_3, u and r as functions of the initial parameters r_0, u_0 and Λ. We confine ourselves to quoting two of these conditions,

$$\Gamma'^{-1}(k = 0) = r, \qquad \frac{\partial}{\partial k^2}\Gamma'^{-1}(k)|_{k=0} = 1,$$

which amount to requiring that for small k we have $\Gamma'^{-1} \sim r + k^2$. By means of these manipulations one manages to express the initial correlation function $\Gamma(k;u_0,r_0,\Lambda)$ in the form

$$\Gamma(k; u_0, r_0, \Lambda) \sim Z_3 \Gamma'(k; u, r)$$

What is more important, one manages also to isolate the dependence on Λ, making it possible in the limit $\Lambda \to \infty$ to obtain finite expressions in terms of the so-called physical parameters r and u. For dimensional reasons, Z_1 and Z_2 are functions only of u (and of $\varepsilon = 4 - d$).

The Callan–Symanzik equations are derived by expressing a trivial truth, namely that the initial correlation function Γ cannot depend on the formal manipulations carried out subsequently; in this way one obtains an equation of the form

$$\left[m\frac{\partial}{\partial m} + \beta(u)\frac{\partial}{\partial u} - \gamma_3(u) \right]\Gamma'^{-1}(k; m, u) = \text{R.H.S.} \tag{5.12}$$

where $m^2 = r$, and

$$\beta(u) = -\varepsilon\left[\frac{\partial}{\partial u} \log(uZ_1(u)Z_3^{-2}(u)) \right]^{-1}$$

$$\gamma_3(u) = \beta(u)\frac{\partial}{\partial u}\log Z_3(u)$$

In view of their provenance, equations like (5.12) are *a priori* empty; nevertheless one can show, and this is the important point, that when $d = 4$ the right-hand-side of (5.12) becomes negligible for large momenta $k \gg m$. The argument then proceeds on physical grounds as follows: in the critical domain ($k, m \ll \Lambda$) the coupling constant u_0 can be considered to be infinite, because for dimensional reasons it is of order Λ^ε; but by virtue of (5.11),

$$u_0^{1/\varepsilon} = m\exp\left[-\int^u \frac{du'}{\beta(u')} \right].$$

The criticality condition, that u_0 become infinite, can be satisfied if $\beta(u)$ has a zero, say at u^*; near such a zero one can write

$$\beta(u) \sim \omega(u - u^*),$$

whence

$$u_0^{1/\varepsilon} \sim m\exp\left[-\frac{1}{\omega}\log|u - u^*| \right].$$

For u_0 to become infinite at $u = u^*$ it is necessary further that ω be positive; one says then that u^* is an attractive infrared zero. When $\omega < 0$ one says that u^* is an attractive ultraviolet zero; this second alternative bears on the behaviour at $k \gg \Lambda$, which is the problem usually considered in field theory (see Figure 5.7).

The coefficient ω which governs the variation of $\beta(u)$ near the fixed point u^*, thereby governs the corrections to scaling; indeed ω is nothing but the exponent

$-y_{\rm II}$ of Section 5.3. From (5.12) we see that the exponent η is given by

$$\eta = \gamma_3(u^*)$$

In the same way one can determine all the critical exponents, scaling functions and so on.

Accordingly, the first step in this approach is to calculate the function $\beta(u)$; near $d = 4$ one finds

$$\beta(u) = -\varepsilon u + \frac{(n + 8)}{6}u^2 + \cdots \tag{5.13}$$

This shows that $\beta(u)$ does have a non-trivial zero, $u^* \sim \varepsilon$, which is attractive in the infrared (see Figure 5.8). In Figures like (5.7) and (5.8) one could draw arrows to indicate which points are attractive and which repulsive; the direction of the arrows is given by

$$\frac{du}{dl} \sim \mp \beta(u)$$

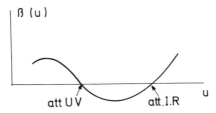

Figure 5.7.   A typical curve $\beta(u)$ having an 'attractive ultraviolet' fixed point (negative slope), and an 'attractive infra-red' fixed point (positive slope)

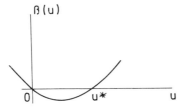

Figure 5.8.   General appearance of the curve $\beta(u)$ as obtained from an expansion in powers of $\varepsilon = 4 - d$

the negative sign being appropriate to the problems of statistical mechanics (in the infrared), and the positive sign to those of field theory (in the ultraviolet). The marginal case arises when $\beta(u)$ touches the horizontal axis, which it does for $\varepsilon = 0$, as shown by (5.13).

Finally we stress some features of this approach along the lines of the first version: we have not introduced new couplings; we have allowed the cutoff parameters to tend to infinity; and the value of u^* is universal. By contrast, in the second version, the renormalization does introduce new couplings; the cutoff parameter is maintained at its initial value; and the positions of the fixed points are not universal. This comparison would tend to incline one to the first version; however, in the first version one loses the physics of the systematic reduction in the number of degrees of freedom, and the topological description in parameter space.

5.7 Conclusion

The expansions in powers of $\varepsilon = 4 - d$ constitute an outstanding achievement in physics. They have revolutionized our methods for approaching problems and our assessment of approximation procedures. They have contributed to reformulating the questions we ask. They are a major challenge, to be pondered in depth.

References

Jona-Lasinio, G. (1973), in *Collective Properties of Physical Systems*, Nobel Symposium XXIV, Academic Press.
Larkin, A. I., Khmelnitskii, D. E. (1969), *Sov. Phys. J.E.T.P.*, **29,** 1123.
Roulet, B., Gavoret, J., Nozières, P. (1969), *Phys. Rev.*, **178,** 1072.
Wegner, F. J., Houghton, A. (1973), *Phys. Rev. A*, **8,** 401.
Wegner, F. J. (1974), *J. Phys. C*, **7,** 2098.
Wilson, K. G. (1971), *Phys. Rev. B*, **4,** 3174, 3184.
Wilson, K. G. (1972), *Phys. Rev. Letters*, **28,** 548.
Wilson, K. G., Fisher, M. E. (1972), *Phys. Rev. Letters*, **28,** 240.
Wilson, K. G., Kogut, J. (1974), *Phys. Reports* 12 *C*, No. 2.

CHAPTER 6
Results on simple systems

> '*I mean the art of arranging a large number of objects in systematic fashion, in a way that enables one to see their interrelations at a glance, to understand quickly how they combine, and to form new combinations oneself. We shall demonstrate the usefulness of this art which, though still in its infancy, will, when fully developed, offer both the advantage of condensing into a small table what it would otherwise be difficult to explain in a large book; and also the means, more valuable still, of presenting isolated facts in the way best adapted for drawing general conclusions from them.* Condorcet

The present chapter aims to display in an orderly fashion the results so far obtained for simple systems. The distinction between simple and complex (or composite) systems is central to the organization of this book. Complex systems differ from simple ones by the presence of one or more perturbations. The possible perturbations are so varied in nature than an enumeration would be very tedious; the chapters which follow will give a fairly clear picture of what is involved.

Within the Ginzburg–Landau formulation as described in Chapter 2, simple systems are conveniently described by the simplest of the standard expressions for the free-energy density; this is invariant under rotations in real space and in the space spanned by the order parameter, contains forms quadratic and quartic in the local field, and also a form quadratic in its gradient.

The critical properties of simple systems depend only on the dimensionality d of real space and on the dimensionality n of the space spanned by the order parameter. This remarkable universality admits a Mendeleeff-type classification of simple systems by tabulation according to n and d.

The main results relate to the critical exponents, the equation of state and the correlation function. The results of the renormalization-group approach, and in particular of the expansion in powers of $\varepsilon = 4 - d$ as described in the last chapter, can then be fitted into the general context of results obtained by whatever means.

6.1 Tabulation by n and d

Ignoring the values $n < -2$ and $d < 0$ whose relevance is doubtful, we distinguish three main regions of the (n, d) table shown in Figure 6.1. First, the classical region $d > 4$, where exponents take their classical values. Second, the

100

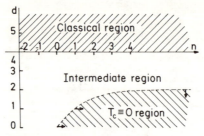

Figure 6.1. (n, d) tabulation showing the three main regions. The boundary between the intermediate and the $T_c = 0$ regions is not known exactly, but it is known to pass through the points $(n = 0, d = 0)$, $(n = 1, d = 1)$, $(n = \infty, d = 2)$. The arrows point from these towards the interior of the $T_c = 0$ region

region $T_c \equiv 0$, where singularities at finite temperatures are forbidden, so that the only singular temperature is absolute zero. Third, the intermediate region, which is of course the region of greatest interest as regards critical phenomena.

The classical and the intermediate regions are separated by the line $d = 4$ which was studied in Section 5.4. The boundary between the intermediate and the $T_c \equiv 0$ regions is less well explored; we know only three points on it, namely the point $(n = d = 0)$, the point $(n = d = 1)$ and the point $(n = \infty, d = 2)$. It would be interesting to locate the line in full; this remains an open problem.

As regards the region $T_c \equiv 0$, we should recall that the results here have only a low degree of universality, as discussed at the end of Section 2.1; and in any case, even if one wished to extend the definition of critical exponents into this region, one is forced into some redefinitions by the importance which now attaches to the various factors $\beta = 1/T$. All in all the region $T_c \equiv 0$ is too special to be treated here.

We shall see in the next subsection that inside the intermediate region the critical exponents appear to vary in a regular manner. But one should note the special effects of gauge variables, $(n > 1)$, on some critical properties. Thus, in the presence of gauge variables, the transverse susceptibility remains infinite in the low-temperature phase, since an arbitrarily small transverse field suffices to rotate the spontaneous magnetization en bloc. This entails a whole series of consequences which we shall have to describe later. In particular, for $d < 2$ there can be no spontaneous magnetization when gauge variables are present. Consequently the semi-infinite line $(d = 2, n > 1)$ plays a very special role. While for $(n = 1, d = 2)$ the solution of the two-dimensional Ising model shows that one has a normal phase transition, for $(n > 1, d = 2)$ it seems that there is still a phase transition but of a rather special type; this problem of short long-range order is discussed in Section 13.2.

Before discussing the results having universal validity one should say a word about the variation of the critical temperature, whose magnitude is known to be not at all universal. Nevertheless for a given model with given coupling constants it makes sense to enquire about the variation of T_c across the (n, d) table. By taking into account the location of the region $T_c \equiv 0$, the results for $n = -2$ and $n = \infty$, plus those for $d = 1, d = 0$, etc., one can readily reach the following conclusions: the critical temperature decreases when, in the (n, d) table, one moves from left to right (d constant, n increasing), or from top to bottom (n constant, d decreasing). It seems that for $n > 2$, $d = 2$, T_c is identically zero, (Brézin, Zinn-Justin, 1976; Migdal, 1975).

6.2 Critical exponents

In view of the scaling laws, the main critical exponents are determined by any two of them. It is entirely arbitrary which two are chosen; here we select η and γ.

Expansion in power of ε

Within the framework of power-series expansions in ε, the exponent η has been calculated up to order ε^4, and the exponent γ up to order ε^3; one finds (Brézin, Le Gouillou, Zinn-Justin, Nickel, 1973)

$$\eta = \varepsilon^2 \frac{(n + 2)}{2(n + 8)^2} \left\{ 1 + \varepsilon \left[\frac{6(3n + 14)}{(n + 8)^2} - \frac{1}{4} \right] + \varepsilon^2 \left[\frac{45(3n + 14)^2}{(n + 8)^4} \right. \right.$$

$$\left. \left. - \frac{8(3n^2 + 53n + 160)}{(n + 8)^3} + \frac{(-5n^2 + 234n + 1076)}{16(n + 8)^2} - \frac{24(5n + 22)}{(n + 8)^3} \cdot \zeta(3) \right] \right\}$$

$$+ O(\varepsilon^5) \tag{6.1}$$

$$\gamma = 1 + \varepsilon \frac{(n + 2)}{2(n + 8)} + \varepsilon^2 \frac{(n + 2)(n^2 + 22n + 52)}{4(n + 8)^3}$$

$$+ \varepsilon^3(n + 2) \left[\frac{55n^2 + 268n + 424}{2(n + 8)^5} + \frac{3(n + 2)(n + 3)}{(n + 8)^4} + \frac{(n + 2)^2}{8(n + 8)^3} \right.$$

$$\left. - \frac{6(5n + 22)}{(n + 8)^4} \cdot \zeta(3) \right] + O(\varepsilon^4) \tag{6.2}$$

Notice the rapidly increasing complication of the higher order terms and the appearance of the coefficient $\zeta(3) \approx 0\cdot60103$; ($\zeta(x)$ is Riemann's zeta-function). The presence of such a transcendental coefficient is a severe blow to 'Daltonian' prejudices, which would prefer critical exponents in physical cases to be simple rational fractions.

If for $\varepsilon = 1$ we compare these values of η and γ to the results of numerical calculations for three dimensions, with $n = 1$ and $n = 3$, we find that the last terms displayed in the expansions cause the agreement with the expected values to become worse instead of better. In fact it is suspected that as a rule the series are asymptotic but not convergent (though they do converge for $n = -2$ and for n very large). Asymptotic series abound in physics; and in the present instance the qualitative difference between what happens for $\varepsilon > 0$ and for $\varepsilon < 0$ lends support to the suspicion that the expansions are asymptotic: (recall Dyson's (1952) argument on the convergence of the perturbation series in quantum electrodynamics).

For $n = 1$ and $d = 3$ the result of numerical calculations for γ is

$$\gamma \approx 1{,}250 \pm 0{,}003;$$

for comparison, the expression (6.2) yields $\gamma \approx 1{,}244$ if one stops with the ε^2 term, and $\gamma = 1{,}195$ if one stops with the ε^3 term. This suggests that the optimal value is obtained if one stops the expansion at a fairly low order; in the present case with the term of second order. Under such conditions the calculation of higher-order terms, whose results are in any case unattractively complicated, is of interest only if one proposes to exploit special tricks like Padé approximants for expediting the summation (see also the discussion in Section 6.5).

Expansion on powers of $1/n$

We proceed to the result of power-series expansions in $1/n$. The exponents η and γ are available to first order:

$$\eta = \frac{\varepsilon^2}{n} \cdot \left(\frac{2 - \varepsilon}{4 - \varepsilon}\right) \cdot A(\varepsilon) + O\left(\frac{1}{n^2}\right) \tag{6.3}$$

$$\gamma = \frac{2}{2 - \varepsilon} - \frac{3\varepsilon}{n} \cdot A(\varepsilon) + O\left(\frac{1}{n^2}\right) \tag{6.4}$$

where $A(\varepsilon)$ can be expressed in terms of gamma functions:

$$A(\varepsilon) = \left(\frac{\sin \pi\varepsilon/2}{\pi\varepsilon/2}\right) \cdot \frac{\Gamma(2 - \varepsilon)}{\Gamma^2\left(2 - \frac{\varepsilon}{2}\right)}$$

Notice that $A(\varepsilon) = 1 - \dfrac{\varepsilon^2}{4} + O(\varepsilon^3)$; $A(1) = \dfrac{8}{\pi^2}$; $A(2) = \dfrac{1}{2}$. One can check that this expansion in $1/n$ is consistent with that in ε in the region where the two overlap (ε small, n large).

The expressions above cannot be used as they stand for estimating exponents at physical points like $(\varepsilon = 1, n = 1)$ or $(\varepsilon = 1, n = 3)$. The expansions are too short for this; they are too short also to give any idea whether the series converge. Nevertheless the expansion in $1/n$ is very valuable for testing a variety of hypotheses.

Other results

In order to push the (n, d) tabulation as far as possible, one should add to the above results some others that are exact, or stem from numerical calculations.

Exact results include those for ($n = -2$, d arbitrary), (see Section 2.6):

$$\eta = 0, \gamma = 1$$

those for ($d = 1$, $-2 < n < 1$), (Balian and Toulouse, 1974):

$$\eta = 1, \gamma = 1$$

and the results for ($n = 1$, $d = 2$), (the Onsager solution):

$$\eta = \tfrac{1}{4}, \gamma = \tfrac{7}{4}$$

There exist plausibility arguments (see Section 9.3), though no rigorous proof, suggesting that for ($n = 2, d = 2$) one has

$$\eta = \tfrac{1}{4}.$$

From amongst numerical results we quote the following values obtained for $n = 0$:

for $d = 3$: $\eta \sim \tfrac{1}{18}, \quad \gamma \sim \tfrac{7}{6}$

for $d = 2$: $\eta \sim \tfrac{2}{9}, \quad \gamma \sim \tfrac{4}{3}$

and the following values obtained for three dimensions, $d = 3$:

for $n = 1$: $\eta = 0\!\cdot\!04 \pm 0\!\cdot\!01, \quad \gamma = 1\!\cdot\!250 \pm 0\!\cdot\!003$

for $n = 3$: $\eta = 0\!\cdot\!04 \pm 0\!\cdot\!01, \quad \gamma = 1\!\cdot\!37 \pm 0\!\cdot\!02$

Trends in the (n, d) table

By taking all the above results into account one can plot curves of constant η, constant α, etc., through the intermediate region of the (n, d) plane. This framework of curves provides the most practical method of interpolation, and thus suggests some plausible predictions.

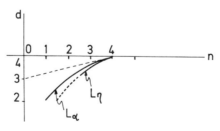

Figure 6.2. Sketch of the curves L_α and L_η in the plane (n, d). Both lines pass through the point $(n = d = 4)$, and have a common tangent there (the broken line). One knows, further, that the curve L_α passes through the point $(n = 1, d = 2)$. The broken curve extrapolating L_η is conjectural (see Chapter 9)

Here we confine ourselves to comments on some special features. One notices that when d is constant and n increases from -2 to infinity, the exponents γ and v increase, the exponent α decreases and changes size, the exponent η goes through a maximum, and the exponent δ, which is directly related to η through

$$\delta = \frac{d + 2 - \eta}{d - 2 + \eta},$$ goes through a minimum. Thus the intermediate region of the

(n, d) plane contains two noteworthy curves, shown in Figure 6.2: the locus L_α of points where α is zero, and the locus L_η of point where η has a maximum under variation of n with d kept constant. The equation for L_α is obtained in the form of a power-series expansion in ε:

$$n = 4 - 4\varepsilon + \frac{\varepsilon^2}{2}(1 + \zeta(3)) + O(\varepsilon^3) \tag{6.5}$$

and the equation for L_η is

$$n = 4 - 4\varepsilon + \frac{\varepsilon^2}{2}(\tfrac{5}{3} + 11\zeta(3)) + O(\varepsilon^3) \tag{6.6}$$

Notice that the two curves touch at ($n = d = 4$). One knows moreover that L_α passes through the point ($n = 1, d = 2$); while L_η lies below L_α (at least for small ε) and seems to lie between (not through) ($n = 1, d = 2$) and ($n = 2, d = 2$).

Other exponents

Beside the main critical exponents there are also others. Thus we see, referring back to Section 5.3, that the exponent v follows from y_1, the larger of the two eigenvalues associated with the non-trivial fixed point, while the smaller eigenvalue y_{II} governs the corrections to the asymptotically dominant terms. One has

$$g_1 \sim s^{y_1} \quad \text{where} \quad y_1 = \frac{1}{v} > 0$$

$$g_{II} \sim s^{y_{II}} \quad \text{where} \quad y_{II} < 0$$

Thus the irrelevant scaling field g_{II} multiplies the asymptotically dominant terms by the correction factor

$$1 + g_{II}(\Delta T)^{-vy_{II}}$$

Accordingly, the susceptibility χ may be written as

$$\chi \sim (\Delta T)^{-\gamma}[1 + g_{II} \cdot (\Delta T)^{-vy_{II}}] \tag{6.7}$$

Within the framework of the expansions in powers of ε and $1/n$, y_{II} is determined by

$$y_{\mathrm{II}} = -\varepsilon + \frac{3(3n + 14)}{(n + 8)^2} \cdot \varepsilon^2$$

$$-\left[\frac{33}{4}n^2 + \frac{461}{2}n + 740 + 24(5n + 22)\zeta(3) - \frac{18(3n + 14)^2}{n + 8}\right]\frac{\varepsilon^3}{(n + 8)^3}$$

$$+ O(\varepsilon^4) \tag{6.8}$$

$$y_{\mathrm{II}} = -\varepsilon + \frac{\varepsilon^2}{n} \cdot \frac{(2 - \varepsilon)(3 - \varepsilon)^2}{2(4 - \varepsilon)} \cdot A(\varepsilon) + O\left(\frac{1}{n^2}\right) \tag{6.9}$$

The following notation is widespread:

$$y_{\mathrm{I}} = d - d_{\phi^2}, \qquad y_{\mathrm{II}} = d - d_{\phi^4}$$

where d_{ϕ^2} and d_{ϕ^4} are the anomalous dimensions of the scaling operators conjugate to the fields g_{I} and g_{II}.

6.3 The equation of state

We saw in Section 3.1 that in the critical region ($H \to 0$, $T \to T_c$) the equation of state $H = f(M, T)$ takes the homogeneous form

$$\frac{H}{M^\delta} = f\left(\frac{t}{M^{1/\beta}}\right) \tag{6.10}$$

The scaling function $f(x)$ has the same universality class as the critical exponents; for simple systems it is fully specified by n and by d. Before presenting detailed results we recall some general properties of $f(x)$.

General properties of the scaling function

First of all, the scaling function $f(x)$ is normalized by the following two conditions: above the critical point and in zero field, $M \sim (-t)^\beta$; while below the critical point, $H \sim M^\delta$. Thus $f(x)$ is defined only for $-1 < x < \infty$, and the curve $y = f(x)$ passes through the points $(-1, 0)$ and $(0, 1)$. The branches for $x < 0$ and $x > 0$ correspond, respectively, to the low-temperature and the high-temperature phases. In the neighbourhood of the critical isotherm $x = 0$ the function $f(x)$ is analytic (regular). The points $x = 1$ and $x = \infty$ correspond to the critical isochore $H = 0$, the former when $T < T_c$ and the latter when $T > T_c$. In the limit $x \to \infty$, or in other words for $T > T_c$ and $H \to 0$, $M \to 0$, the analyticity and homogeneity properties of the equation of state imply the following expansion of $f(x)$ (Griffiths, 1967):

$$f(x) \sim \sum_{l=1} a_l x^{-(2l-1-\delta)\beta} \sim a_1 x^{\gamma} + a_2 x^{\gamma-2\beta} + \cdots \tag{6.11}$$

Accordingly, the function $f(x)$ is positive and increases monotonically, vanishes for $x = -1$, passes through 1 at $x = 0$, and for large x behaves like x^{γ}. In Landau theory it takes very simple form

$$f(x) = 1 + x \tag{6.12}$$

Expansion in powers of ε

As a power series in ε, for $\varepsilon > 0$, the scaling function $f(x)$ has been calculated to order ε^2 (Brézin, Wallace, Wilson, 1973); one finds

$$f(x) = 1 + x + \varepsilon f_1(x) + \varepsilon^2 f_2(x) + O(\varepsilon^3) \tag{6.13}$$

where $f_1(x)$ and $f_2(x)$ are fairly complicated functions. For instance, $f_1(x)$ is given by

$$f_1(x) = \frac{1}{2(n+8)} \{3(x+3)\ln(x+3) + (n-1)(x+1)\ln(x+1) + 6x\ln2$$

$$- 9(x+1)\ln3\} \tag{6.14}$$

As x tends to infinity, one recovers the singular behaviour (6.11):

$$f(x) \sim x\left(1 + \frac{\varepsilon(n+2)}{2(n+8)}\ln x + \cdots\right) \sim x^{\gamma} \tag{6.15}$$

with γ given by the ε-expansion indicated in Section 6.2.

When there are gauge variables, (for $n > 1$), we must distinguish between the longitudinal and transverse suceptibilities χ_L and χ_T; for $T < T_c$, both diverge when one approaches the coexistence curve by letting H tend to zero. While χ_T diverges like H^{-1}, calculations to order ε^2 and some exact results to be discussed later both suggest (Brézin, Wallace, 1973) that χ_L diverges like

$$\chi_L \sim H^{-\varepsilon/2} \tag{6.16}$$

Expansion in powers of $1/n$

In the limit $n = \infty$, which corresponds for the spherical model studied in Section 2.6, one finds for $2 < d < 4$:

$$f(x) = (1 + x)^{2/(d-2)} = (1 + x)^{\gamma} \tag{6.17}$$

For $d = 4$ the equation of state is

$$H \sim M(t+M)^2[\ln(t+M^2)]^{-1} \tag{6.18}$$

It cannot in this case be expressed in truly homogeneous form. Finally, the assumption (6.16) about the divergence of χ_L is satisfied in the limit $n = \infty$.

The function $f(x)$ has been calculated to first order in $1/n$ by Brézin and Wallace (1973):

$$f(x)^{1/\gamma} = 1 + x + \frac{1}{n}[g(x) - (1 + x)g(0)] \qquad (6.19)$$

where $g(x)$ is given by a fairly complicated integral, and γ by the $1/n$-expansion indicated in Section 6.2.

Other results

In the Gaussian model one cannot speak of an equation of state because the low-temperature phase is not well defined. By contrast, a perfectly ordinary equation of state does exist in the limiting case $n = -2$ which has the same main critical exponents as the Gaussian model. The two-dimensional Ising model, whose critical exponents are known exactly, has not been solved in an external field, and its equation of state is unknown.

Balian and Toulouse (1974) have found the equations of state for $d = 0$ and $-2 < n < 0$, and for $d = 1$ and $-2 < n < 1$, and these can be written in homogeneous form.

In order to explore the universality class of the scaling function $f(x)$, Milosevic and Stanley (1972) have used numerical analysis of high-temperature series expansions to determine the equation of state for the Heisenberg model ($n = 3$), with various magnitudes of the quantized spin and on various cubic lattices ($d = 3$); the same has been done for the Ising model ($n = 1$), which has also been tackled with low-temperature series. Some numerical results for the Ising model are shown in Figure 6.3 which plots $f(x)$ against x for $(1 + x)$ between 0 and 5. The numerical estimates agree very well, especially for the Ising model, with the results of the ε-expansion (when ε is set equal to 1 as for $d = 3$), and the agreement improves on including terms of higher order in ε (agreement is within 3.10^{-3} for $x < 0$ and within 10^{-3} for $x > 0$). One expects that to order ε^3 the gap would widen.

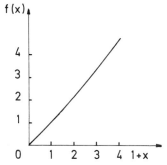

Figure 6.3.   The scaling function for the Ising model on a cubic lattice ($n = 1$, $d = 3$) from numerical calculations

108

The scaling function can also be obtained directly from experiment: see Figure 3.1 and the discussion in Section 3.1.

Corrections to asymptotic homogeneity

Only in the asymptotic limit $t \to 0$, $H \to 0$ does the equation of state take a homogeneous form. The first departures from homogeneity are due to the irrelevant field g_{II} with negative dimension y_{II}, introduced at the end of Section 6.2. One must distinguish the two cases where $x = \left(\dfrac{t}{M^{1/\beta}} \right)$ is large or small; for large x, the correction term in the equation of state involves $|t|^{|y_{II}|/y_{I}}$ and can be written as

$$\frac{H}{M^{\delta}} = f(x)\{1 + \text{constant} \cdot t^{|y_{II}|/y_{I}}\} \tag{6.20}$$

while for small x, i.e. near the critical point, one has

$$\frac{H}{M^{\delta}} = 1 + \text{constant} \cdot M^{|y_{II}|/\beta y_{I}} \tag{6.21}$$

The spherical model ($n = \infty$) provides a remarkable illustration of these corrections; for the spherical model on a cubic lattice (Joyce, 1972) the equation of state takes the form

$$\frac{H}{M} = \sum_{n=0} \omega_n \left(\frac{t + M^2}{1 + t} \right)^{n+2} \tag{6.22}$$

where the ω_n are certain numbers. This reproduces exactly the form of the corrections (6.20) and (6.21) with $y_{II}/y_{I} = -1$; in fact for the spherical model one has $y_{I} = d - 2$ and $y_{II} = d - 4$.

6.4 The correlation function

We shall consider the two-point correlation function

$$\Gamma(R) = \langle M(0)M(R) \rangle - \langle M(0) \rangle \langle M(R) \rangle,$$

and its Fourier transform $\Gamma(k)$, in the critical region $\Delta T \to 0$ and $H \to 0$, concentrating mainly on the high-temperature phase $\Delta T > 0$ in low field.

General properties of the scaling function

Near the critical point the correlation function $\Gamma(k, \Delta T)$ in zero field ($H = 0$) takes for $k \ll 1/a$ the homogeneous form already introduced in Section 3.1,

$$\Gamma(k, \Delta T) = \frac{1}{k^{2-\eta}} g(x) \tag{6.23}$$

where $x = k(\Delta T)^{-\nu}$.

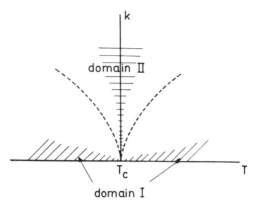

Figure 6.4. Regions I ($k\xi \ll 1$) and II ($k\xi \gg 1$) of the (k, T) plane. The broken lines indicate where $k\xi \sim 1$ [i.e. $k \sim (\Delta T)^\nu$]

The scaling function $g(x)$ is defined for x between zero and infinity. The limit $x = 0$ is reached by letting k tend to zero with ΔT small and fixed. In the neighbourhood of $k = 0$, $\Gamma(k, \Delta T)$ is analytic and can be written

$$\Gamma(k, \Delta T) = \Gamma(0, \Delta T)D(k^2\xi^2) \sim \Delta T^{-\gamma}(b_0 + b_1.(k^2\xi^2) + b_2(k^4\xi^4) + \cdots)$$
(6.24)

which implies the following expansion of $g(x)$:

$$g(x) \underset{(x \to 0)}{=} \sum_{l=0} a_l x^{2 - \eta + 2l}$$
(6.25)

The other limit $x \to \infty$ is reached by letting ΔT tend to zero while $k \neq 0$. It relates to the behaviour of the correlation function at distances short compared to ξ, though long compared to the range of forces. In this regime there are contributions with singularities like $(\Delta T)^{1-x}$, which will be discussed later. Figure 6.4 shows the two regions I and II of the $(k, \Delta T)$ plane corresponding to the correlation function taken at long distances ($k\xi \ll 1$) and short distances ($k\xi \geqslant 1$), respectively.

In a field H the expansion (6.24) for small k is generalized as follows:

$$\Gamma(k, \Delta T, H) = \chi(\Delta T, H)D(k^2\xi^2(\Delta T, H))$$
(6.26)

By contrast, in the limit of large $k\xi$ it is no longer possible to express the correlation function Γ as a homogeneous form of the same kind.

In Landau theory, as in the spherical and Gaussian models and as for $n = -2$, the scaling function $g(x)$ takes the very simple Ornstein–Zernike form

$$g(x) = \frac{1}{1 + x^{-2}}$$
(6.27)

Expansion with singularities (The correlation function at short distances)

Consider the situation where $T > T_c$ and $H = 0$. When $k\xi \geqslant 1$, i.e. in region II, or in other words when $R \ll \xi$ in real space, the short-distance correlation function involves the product $M(r_1)M(r_2)$ of two local variables at near-neighbouring points. According to the reduction hypothesis of Section 4.4 this product can be expanded as a linear combination of local scaling operators $O_i(\mathbf{k})$ taken in increasing order of their anomalous dimensions x_i. We can write this expansion as

$$M(\mathbf{r}_1)M(\mathbf{r}_2) = \sum_i A_i(\mathbf{R})\, O_i(\boldsymbol{\rho})$$

where   (6.28)

$$\mathbf{R} = \mathbf{r}_1 - \mathbf{r}_2 \quad \text{and} \quad \boldsymbol{\rho} = \tfrac{1}{2}(\mathbf{r}_1 + \mathbf{r}_2)$$

Accordingly, when Γ is considered as a function of ΔT for fixed k, one expects it to contain terms with singularities like $(\Delta T^{1/y_1})^{x_i}$, generated by the operators in the above expansion, and appearing as factors multiplying other, non-singular, expressions involving $(\Delta T)^m$. The lowest dimensionalities belong to the unit operator, to O_I, and to O_{II}; they are, respectively, $x_0 = 0$, $x_I = d_\phi^2 = d - y_I$, and $x_{II} = d_{\phi^4} = d - y_{II}$. Remembering that $y_I = 1/v$, and the relation $vd = 2 - \alpha$, which hold for $d < 4$, it follows that for $d < 4$ the expansion in powers of ΔT becomes

$$\Gamma(\mathbf{k}, \Delta T) = A_1(k) + A_2(k)\Delta T + A_3(k)\Delta T^{1-\alpha} + A_4(k)\Delta T^2 + A_5(k)\Delta T^{2-\alpha}$$

$$+ A_6(k)\Delta T^{2-\alpha-y_{II}v}$$ (6.29)

For the scaling function $g(x)$ this implies the singular expansion

$$g(x) = C_1 + C_2 x^{-1/v} + C_3 x^{-(1/v)(1-\alpha)} + C_4 x^{-2/v}$$

$$+ C_5 x^{-(1/v)(2-\alpha)} + C_6 x^{-(1/v)(2-\alpha)+y_{II}} + \cdots$$ (6.30)

where the universal constants, C_1, C_2, C_3 can in principle be expressed as functions of n and of d.

Brézin, Le Gouillou and Zinn-Justin (1974a) have obtained a similar expansion in the more general case where $T \gtrsim T_c$ and $H \neq 0$; it involves, indirectly, the scaling function $f(x)$ from the equation of state, and for $n > 1$ contains a new term stemming from the polarization of the order parameter, and proportional to $(\Delta T)^{2-\alpha-\phi}$; ($\phi$ is the crossover exponent for quadratic anisotropy, (chapter 8), linked directly to the anomalous dimension of the operator M_iM_j, $i \neq j$).

Expansion in powers of ε

In the expansion (6.24) of $\Gamma(k, \Delta T)$ near $k = 0$, the coefficients $b_1, b_2, \ldots$ have been obtained as power series in ε (for arbitrary n), and show that corrections to the Ornstein–Zernike expression

$$\Gamma(k, \Delta T) \underset{(k \to 0)}{\sim} \frac{\Gamma(0)}{1 + k^2 \xi^2}$$

enter only to order ε^2; consequently such corrections are small, as had already been found in numerical work (Fisher, Burford, 1967).

In the expansion (6.30) of $g(x)$ for large x, the coefficients C_1, C_2, C_3 have been obtained to order ε^2 (Fisher, Aharony, 1973). To order ε they become

$$C_1 = 1 + O(\varepsilon^2)$$

$$C_2 = \frac{-6}{4 - n} - \frac{(n + 2)(7n + 20)}{(n + 8)(4 - n)^2} \varepsilon + O(\varepsilon^2) \tag{6.31}$$

$$C_3 = \frac{n + 2}{4 - n} + \frac{(n + 2)(7n + 20)}{(n + 8)(4 - n)^2} \varepsilon + O(\varepsilon^2)$$

For $n = -2$, $n \to \infty$ and $\varepsilon = 0$ one recovers, as expected, the beginnings of an Ornstein–Zernicke type of expansion.

Note that in the limit $x \to \infty$, one can express $g(x)$ in the form

$$g(x) = 1 - x^{-1/\nu} + (\gamma - 1)x^{-1/\nu}\left(\frac{x^{\alpha/\nu} - 1}{\alpha}\right) + O(\varepsilon^3) + \text{singular terms} \tag{6.32}$$

where γ and α are understood to be replaced by their ε-expansions to order ε^2, as given in Section 6.2. We see that for $n = 4 + O(\varepsilon)$, α vanishes and the coefficients C_2 and C_3 diverge. From (6.32), $g(x)$ can be written

$$g(x) = 1 - x^{-1/\nu} + (\gamma - 1)x^{-1/\nu}[\log(x^{1/\nu})] + \cdots \tag{6.33}$$

Let C_3' be the coefficient of the singularity $|\Delta T|^{1-\alpha}$ for $\Delta T < 0$; then up to sign the ratio C_3/C_3' is equal to the ratio A^+/A^- between the coefficients appearing in the specific heat above $(+)$ and below $(-)$ T_c;$(C \sim A^{\pm}|\Delta T|^{-\alpha})$. This equality hinges on the fact that both the singular term involving $(\Delta T)^{1-\alpha}$ and the specific heat are governed by the energy operator. The ratio in question has been calculated to order ε^2; so has the analogous ratio C^+/C^- for the susceptibility $(\chi = \Gamma(k = 0) \sim C^{\pm}|\Delta T|^{-\gamma})$. They are given by Brézin, Le Gouillou, Zinn-Justin, 1947b)

$$\frac{A^+}{A^-} = -\frac{C_3}{C_3'} = 2^\alpha(1 + \varepsilon)\frac{n}{4} + O(\varepsilon^3),$$

$$\frac{C^+}{C^-} = 2^{\gamma-1}\frac{\gamma}{\beta} + O(\varepsilon^3) \tag{6.34}$$

where α, β, γ are to be replaced by their expansions up to order ε^2. Note that the ratio of the susceptibilities is defined only for $n = 1$, because for $n > 1$ the susceptibility of the low-temperature phase is strictly infinite.

Expansion in powers of $1/n$

In region I, where $k\xi \to 0$, there are only very small deviations from the Ornstein–Zernike form, which is valid for $n = \infty$. As $k\xi \to 0$, the correlation function Γ can be written, to order $1/n$,

$$\Gamma(k, \Delta T) = \frac{\Gamma(0)}{1 + (k\xi)^2 - \Sigma_4(k\xi)^4} \qquad (6.35)$$

where $\Sigma_4 \sim 10^{-3} \cdot \frac{1}{n}$ for $d = 3$. In region II, where $k\xi \to \infty$, Aharony (1974) has calculated the coefficients C_1, C_2, C_3 numerically to first order $1/n$ for $d = 3$.

Other results

In the two-dimensional Ising model, $(n = 1, d = 2)$ the correlation function $\Gamma(R, \Delta T)$ is well known for $\Delta T \gtrless 0$ and $H = 0$, allowing one to confirm the homogeneity hypotheses and to evaluate the ratio $C^+/C^- \approx 37.7$; (Barouch, McCoy, Wu, 1973). This ratio is very different from its classical value $C^+/C^- = 2$, and is fairly well approximated by substituting in (6.34) the values $\gamma = 7/4$ and $\beta = 1/8$.

The universality of the correlation function with respect to the nature of the lattice and the value of the quantized spin has been tested by numerical work on the Ising model for $T \gtrless T_c$ and $H \neq 0$, (Fisher, Burford, 1967; Tarko, Fisher, 1973), and on the Heisenberg model for $T > T_c$ and $H = 0$ (Ritchie, Fisher, 1972). Figure 6.5 displays the correlation function $\Gamma(k, \Delta T)$ as a function of ΔT for various fixed values of ka, and shows that for fixed k it has a maximum at $\Delta T_M \neq 0$. As k tends to zero, the maximum becomes sharper and ΔT_M vanishes like $k^{1/\nu}$. Such a (rounded) peak is due to the competition, at large k, between the terms with ΔT and $\Delta T^{1-\alpha}$ in the expansion (6.29), and has been observed in resistivity measurements

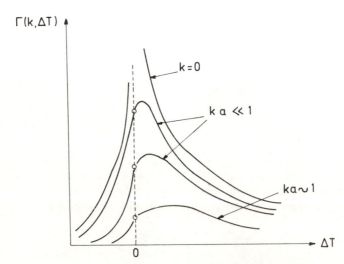

Figure 6.5. Sketch of $\Gamma(k, \Delta T)$ for various values of ka (where a is the lattice spacing)

near the critical point in ferromagnetic metals like iron and nickel. The resistivity anomaly, due to fluctuations of the electron spins, is governed essentially by the temperature dependence of the correlation function $\Gamma(k, \Delta T)$, taken at k values near $2k_F$; (usually the Fermi momentum k_F is small enough for a fairly prominent peak to be observed).

The ratio A^+/A^- of the specific-heat coefficients above and below T_c has been measured (Ahlers, 1973), and Table 6.1 shows the good agreement between the experimental values, and the theoretical values determined from the ε-expansion.

Table 6.1   Values of the ratio A^+/A^-

n	$O(\varepsilon^2)$	experiment
1	0·55	0·53
2	0·99	1·05
3	1·36	1·36

Finally we must mention that the correlation function, like the equation of state, contains correction terms which break the homogeneity, and are due in the first instance to the operator conjugate to the field g_{II}. In the two asymptotic regions I and II the correlation function can be written

$$\Gamma(k, \Delta T) \underset{k\xi \to 0}{=} \Delta T^{-\gamma}\{D(k^2\xi^2) + \Delta T^{-y_{II}/y_{I}} \cdot D_1(k^2\xi^2)\}$$

$$\Gamma(k, \Delta T) \underset{k\xi \to \infty}{=} k^{-2+\eta}\{g(k\xi) + k^{-y_{II}}g_1(k\xi)\}$$

(6.36)

6.5 Discussion of the results

Looking at the results in the last three sections as a whole, one can ask whether some major simplifying features have been missed. The question is particularly acute as regards the dependence on n and d of the exponents, the scaling function, the coefficient ratios, and so on. One could imagine, for instance, that in addition to the four scaling laws there exists another simple relationship between the main exponents; or that some underlying symmetries like invariance under conformal transformations may be important. These are pure speculations, and are mentioned here only to prevent one's awareness and attention from being smothered under the continual accretion of fragmentary results.

Simply for their curiosity value, we shall underline some features suggestive of the existence of a simple approximate theory intermediate between Landau theory and an exact solution, one which would be accurate to the first few orders in ε but not to higher orders, and which would give reasonably good results for $\varepsilon = 1$, or in other words for $d = 3$.

In Section 6.2 we saw that the curves L_α and L_η have a common point ($n = d = 4$) where they touch. In fact the exponents in the (n, d) table do have a symmetry property to first order in η, hence to second order in ε, which generalizes this result,

(Toulouse, 1973). The symmetry can be formulated as follows: to each point (n, d) of the table there corresponds a second point $(n', d' = d)$ given by

$$(n + 2)(n' + 2) = (\bar{n}(d) + 2)^2$$

such that the exponents $\{\alpha'\}$ are obtained from the exponents $\{\alpha\}$ by a transformation T:

$$\{\alpha'\} = T\{\alpha\}$$

defined by

$$(1 - \alpha)(1 - \alpha') = 1, \qquad \eta' = \eta$$

The transformation T will be encountered again in Section 7.3. Notice that $T^2 = 1$. One can check that this symmetry correctly links the two cases with $n = -2$ and $n = \infty$, and ensures that the curves L_α and L_η have the same equation $n = n(d)$. The symmetry is broken by the order ε^3 terms in the exponents, and the breaking is reflected by the difference in curvature between L_α and L_η.

Along the same lines one notes that to the lowest orders in ε (to order ε^2 for $n = 1$), the equation of state can be expressed in a simple parametric form called the linear model. Moreover the ratios of coefficients in (6.34) take very simple forms correct to order ε^2.

It is not clear whether these various properties are all rooted in one and the same approximate theory. If they are, and if one could discover the nature of the approximation and of the terms where it fails, then our understanding of critical phenomena would be considerably enhanced.

6.6 Conclusion

The point worth most emphasis in concluding is undoubtedly the usefulness of the (n, d) table, usefulness of a kind amply commended in the quotation which prefaces this chapter.

References

Aharony, A. (1974), *Phys. Letters*, **46A**, 287.
Ahlers, G. (1973), *Phys. Rev.*, **A8**, 530.
Balian, R., Toulouse, G. (1974), *Ann. Phys.*, **83**, 28.
Barouch, E., McCoy, B. M., Wu, T. T. (1973), *Phys. Rev. Letters*, **31**, 1409.
Brézin, E., Le Gouillou, J. C., Zinn-Justin, J., Nickel, B. G. (1973), *Phys. Letters*, **44A**, 227.
Brézin, E., Le Gouillou, J. C., Zinn-Justin, J. (1974a), *Phys. Rev. Letters*, **32**, 473.
Brézin, E., Le Gouillou, J. C., Zinn-Justin, J. (1974b), *Phys. Letters*, **47A**, 285.
Brézin, E., Wallace, D. J. (1973), *Phys. Rev.*, **B7**, 1967.
Brézin, E., Wallace, D. J., Wilson, K. G. (1973), *Phys. Rev.*, **B7**, 232.
Brézin, E., Zinn-Justin, J. (1976), *Phys. Rev. Letters*, **36**, 691.
Dyson, F. J. (1952), *Phys. Rev.*, **85**, 631.
Fisher, M. E., Aharony, A. (1973), *Phys. Rev. Letters*, **31**, 1238, 1537.
Fisher, M. E., Burford, R. J. (1967), *Phys. Rev.*, **156**, 583.
Griffiths, R. B. (1967), *Phys. Rev.*, **158**, 176.

Joyce, G. S. (1972), in *Phase Transitions and Critical Phenomena*, Vol 2, Academic Press.

Migdal, A. A. (1975), *Journal of Experimental and Theoretical Physics*, **69**, 1457.

Milosevic, S., Stanley, H. E. (1972), *Phys. Rev.*, **B6**, 986, 1002.

Ritchie, D. S., Fisher, M. E. (1972), *Phys. Rev.*, **B5**, 2668.

Tarko, H. B., Fisher, M. E. (1973), *Phys. Rev. Letters*, **31**, 926.

Toulouse, G. (1973), in *Collective Properties of Physical Systems*, Nobel Symposium XXIV, Academic Press.

CHAPTER 7
Complex systems

The last chapter collected results on simple systems. Our next task is to understand some possible causes of deviations from such results, and it calls for a classification and study of complex systems.

In order to convey an impression of the range of problems involved, the present chapter surveys many very varied kinds of perturbations. Thus it provides an introduction to the following chapters, which are devoted to a more detailed investigation of certain typical kinds of complexity.

7.1 Types of perturbations

The perturbations which first come to mind are those by 'constant fields'. These arise from changes in intensive variables, applied fields, or coupling constants. In all such cases the results can be understood in terms of an exploration of parameter space, with its domains, its critical surface and its fixed points.

But there exist other kinds of perturbations which fit less easily into this classification. For instance there are systems that are defined not solely by their intensive variables, but also by global constraints on their extensive variables; in thermodynamic language, this is how systems at constant pressure, say, differ from systems at constant volume. Then there are cases where the order parameter is coupled to other degrees of freedom, forcing one to consider several field variables whose fluctuations are not mutually independent. Examples include: magneto-elastic, i.e. spin-phonon coupling (between the magnetization and the position of atoms); electromagnetic interactions in superconductors; and the coupling of nematic-to-smectic fluctuations to the director. Partially related to these examples, there are various impurity effects: mobile impurities (liquid mixtures); frozen impurities (solid alloys); and the effects of inhomogeneities in general.

Finally there are inhomogeneous perturbations arising from finite-size effects (films, grains, semi-infinite media, etc.) with all the various boundary conditions that can obtain.

In real systems some such perturbations are generally present. Therefore it is important from a practical point of view to study their effects, both for making theoretical predictions and for interpreting experimental results. More generally, such study clarifies and considerably enriches the concept of universality.

7.2 Perturbations by constant fields

In terms of the topological concepts developed in Chapters 4 and 5, a system is represented by a curve in parameter space, i.e. by the physical line, which in general intersects the critical surface at one point. The critical behaviour of the system depends on the domain of the critical surface to which this point belongs. In Chapter 5 we tried to explore the domain structure of the critical surface for systems having dimensionality $d = 4 - \varepsilon$. Especially important is one of these domains, the catchment area of the non-trivial fixed point; it is the domain which corresponds to simple systems in the strict sense of the words, i.e. systems whose critical behaviour is governed by the non-trivial fixed point.

Thus, perturbations by constant fields are analysed in two stages. First one decides whether the perturbation forces the system out of its initial domain; if so, then one must determine the properties of the new domain, i.e. whether it corresponds to first-order transitions, or is the catchment area of a new fixed point, etc. Finally one must express the results as homogeneity rules for the various critical variables, and in particular one must describe the crossover behaviour between one critical regime and the other.

Perturbations of the type (m, ℓ)

As an example we consider perturbations which contribute to the free-energy density (i.e. to the 'Hamiltonian') terms like

$$\mu_i O_i(\mathbf{x})$$

with operators O_i expressible as

$$P_m(\mathbf{M}^2) . H_\ell(\mathbf{M})$$

Here, P_m is a polynomial of degree m and H_ℓ is a harmonic polynomial of degree ℓ (Wegner, 1972). Accordingly, the operator O_i, with i specifying (m, ℓ) is a polynomial of overall degree $(2m + \ell)$ in the components of the field variable.

To lowest order in ε these operators $O_{m,\ell}$ are scaling operators defined with respect to the non-trivial fixed point. To first order in ε, the anomalous dimensions $y_{m,\ell}$ of the associated scaling fields $g_{m,\ell}$ are found to be

$$y_{m,\ell} = 4 - 2m - \ell + \varepsilon \left[\left(m + \frac{\ell}{2} - 1 \right) - \frac{1}{n + 8} [m(2m - 2 + n + 2\ell) \right.$$

$$\left. + (2m + \ell)(2m + \ell - 1)] \right] \tag{7.1}$$

Amongst these anomalous dimensions there are some that we have already met. When $\ell = 0$, the associated operators are isotropic (i.e. scalars) in the space spanned by the order parameter. Thus we recognize $y_{1,0}$ as the anomalous dimension denoted by y_t or y_E in previous chapters, and $y_{2,0}$ as the anomalous dimensions previously denoted by y_{II}. When $\ell \neq 0$, one is dealing with operators that are anisotropic in the space spanned by the order parameter, and can recognize $y_{0,1}$ as the anomalous dimension y_h of the field conjugate to the order parameter.

For $m = 0$ and $\ell = 2$ the perturbation is like a quadratic anisotropy; for $m = 0$

and $\ell = 4$, like a cubic anisotropy. Such perturbations, anisotropic in the space spanned by the order parameter, will be studied in detail in Chapters 8 and 9. But already at this stage we can note their anomalous dimensions, with respect to the non-trivial fixed point, to first order in ε:

$$\text{Quadratic anisotropy:} \qquad y = y_{0,2} = 2 - \frac{2\varepsilon}{n+8}$$

$$\text{Cubic anisotropy:} \qquad y = y_{0,4} = \varepsilon \frac{n-4}{n+8}$$

This shows that (at least for small ε) $y_{0,2}$ is positive for all values of n, whence a quadratically anisotropic field is always relevant; by contrast, $y_{0,4}$ can change sign, so that a cubically anisotropic field is relevant for $n > 4$ and irrelevant for $n < 4$. Apart from their practical importance, such perturbations are interesting also as examples of these two different kinds of behaviour.

We digress to note that by setting $n = -2$ in the formula (7.1) for $y_{m,\ell}$ and comparing the result to the Gaussian value, one gets a vivid illustration of the differences between these two cases.

Spatial anisotropy

By spatial anisotropy one means anisotropy in real space, not to be confused with the quadratic and cubic anisotropies discussed above, which refer to the space spanned by the order parameter. The study of spatial anisotropy has great practical importance; it is also an especially simple illustration of what is meant by crossover.

Consider a system consisting of weakly coupled planes; the value of n is arbitrary. At high temperatures one observes a critical regime characteristic of dimensionality $d' = 2$, governed by fluctuations within each plane. As the temperature tends to T_c, the correlation length increases, the coupling between planes becomes important, and one observes a gradual passage to a regime characteristic of dimensionality $d = 3$.

In the language of the renormalization group, one describes this by saying that the ($d = 2$) fixed point is unstable with respect to the coupling R between planes; the stable fixed point is the one for $d = 3$. The crossover behaviour stems from a competition between these two fixed points; a typical trajectory on the critical surface passes close to the $d = 2$ fixed point before converging to the $d = 3$ fixed point. For small R, the crossover temperature T^* is given by

$$R \sim (T^* - T_c)^\phi,$$

where ϕ is not a new exponent; one has $\phi = \gamma(d')$, γ being the critical exponent for the susceptibility in the high-temperature regime, with effective dimensionality d' (in the present case $d' = 2$).

Thus, spatial anisotropy is an especially simply kind of perturbation, in that the crossover exponent is one of the exponents already known, and both competing fixed points are likewise of an already known type.

Long-range forces

Here we are concerned with forces of infinitely long range, like those discussed in Section 2.5. We saw there that such forces can change the characteristic dimensionality d_c and the classical values of the exponents. Similarly, the non-classical values of the exponents are affected when the long-range forces decrease more slowly than the correlations which would arise just from the short-range forces, i.e. when

$$\sigma \leqslant 2 - \eta$$

where η is the critical exponent appropriate to the fixed point for short-range forces. This change of critical regime can be interpreted in terms of an exchange of stabilities between the short-range fixed point and a new, long-range fixed point. For $n = -2$ and $n = \infty$ there exist exact solutions for the long-range critical behaviour; in them one recognizes the exponents for ideal Bose-gas condensation at constant pressure or constant volume, with bosons having kinetic energy $\varepsilon(k) \sim k^{\sigma}$. The general case is discussed in Section 10.2.

Other perturbations

There are many other perturbations in the constant-field category. One example is the dipole–dipole force, which is a special kind of long-range force (having non-constant sign) of great practical importance (see Section 10.3). Another example is the Dzyaloshinskii–Moriya force, which introduces coupled anisotropies in real space and in the space spanned by the order parameter, and which can lead to helical magnetic structures. There are many more.

7.3 Global constraints

Next we consider systems with a global constraint on an extensive variable. The simplest example is the condensation of an ideal Bose gas at constant volume. In this case the critical exponents are those given in Section 2.6 for $n \to \infty$, while for condensation at constant pressure (system in constant field) they are the same as for the Gaussian model (or for $n = -2$). One can check that the critical exponents $\{\tilde{\alpha}\}$ of a system under a global constraint are obtainable from those for the system at constant field, $\{\alpha\}$, by the transformation T,

$$\{\tilde{\alpha}\} = T\{\alpha\}$$

defined by

$$(1 - \alpha)(1 - \tilde{\alpha}) = 1, \qquad \tilde{\eta} = \eta$$

Both sets of exponents satisfy the four scaling laws.

In fact the effects of global constraints are rather general in nature; one calls this Fisher's renormalization (Fisher, 1968). As a rule a global constraint does not change the order of the transition; the transition remains second order but the critical exponents are renormalized by the transformation T, provided the exponent α at constant field is positive. If α is negative, the exponents are unchanged.

The significance of the transformation T becomes clearer if it is applied to the basic exponents d_ϕ and d_{ϕ^2}, the anomalous dimensions of the scaling operators M and M^2, (more precisely, the operators O_1 or O_E). Then T can be written as

$$\tilde{d}_\phi = d_\phi, \qquad \tilde{d}_{\phi^2} + d_{\phi^2} = d \qquad (7.2)$$

More generally, one has the relation

$$x_E + y_E = d$$

where x_E and y_E are the anomalous dimensions of O_E and of its conjugate field g_E, respectively. We see that the transformation T amounts to interchanging the roles of O_E and of g_E:

$$\tilde{x}_E = \tilde{d}_{\phi^2} = y_E \qquad \tilde{y}_E = x_E = d_{\phi^2}$$

This is just what Fisher (1968) proves by a thermodynamic argument. Under the global constraint, one has

$$\Delta \tilde{T} \sim (\Delta \tilde{T})^{1-\alpha}$$

or in other words

$$\tilde{g}_E \sim O_E$$

If α is negative, the correspondence is dominated by the non-singular terms and there is no renormalization. The crossover exponent is just $\phi = \alpha$, which illustrates another aspect of the effects of the sign of α.

We end this rapid survey with an important example of the effects of global constraints. A mixture of two liquids, say a solution of B in A, can be considered either at given chemical potential ($\mu_B - \mu_A$ constant), or at given concentration (constant concentration c of B in A). Then one expects (see also Section 11.3) that the exponents in a transition at constant concentration (i.e. under a global constraint on the number of impurities) will be connected by a Fisher-type renormalization with the exponents observed at constant chemical potential (i.e. at constant irrelevant field). But at high enough concentrations the order of the transition can evidently change; for instance, one knows that for solutions of ^{3}He in ^{4}He the superfluid transition eventually becomes of first order, with the appearance of a tricritical point (see Chapter 12).

7.4 Couplings to other degrees of freedom

Under this heading one is faced with a vast and little-explored field. A Landau-type analysis shows that the coupling can change the order of the transition. The most searching studies, on magneto-elastic couplings, indicate that the results are very sensitive to the initial assumptions. Though we still lack an overall view of such effects, we know enough to be aware of the great diversity that is possible (Section 11.1). Another case that has been studied is the coupling to gauge fields (e.g. of superconductors to the electromagnetic field); here again the coupling can change the order of the transition (Section 11.2).

Stretching a point, perturbations like the quadratic and cubic anisotropies (discussed in Chapters 8 and 9) could also be considered as couplings between different degrees of freedom, namely between different components of the order parameter. These three types of couplings are useful points of reference when interpreting experimental results in general.

As regards the effects of impurities (Section 11.3), it is important to distinguish mobile impurities, discussed in the preceding section, from frozen impurities. For the former, and at low concentrations, one observes a renormalization of the critical exponents, with crossover from the impurity-free to the renormalized regime; with rising concentration the order of the transition eventually changes in a way related to phase-separation effects. By contrast, the effects of frozen impurities depend on the nature of the order parameter, and especially on the sign of the critical exponent α_n (which specifies the divergence of the specific heat in the n-vector model). When α_n is positive there is a new kind of critical behaviour; but when α_n is negative the critical behaviour remains unaffected by the presence of the disorder due to the impurities (Lubensky, 1975). But it seems that for certain kinds of disorder (for $n = 1$, where there are disorder waves) the mathematics of the transition is completely altered, and that one is then faced with a fundamentally more complicated problem (Griffiths, 1969, McCoy, 1972). Note also the connections between the problems of randomly-distributed impurities and the percolation problem discussed in Appendix 2.2 (Essam, 1972).

7.5 Finite-size effects

By finite-size effects one means all those which depend on the geometry of the system under study (Section 10.5). These include the influence of boundaries on bulk properties like critical temperature and crossover behaviour; and also surface effects in the strict sense, like surface contributions to the thermodynamic variables, changes in correlations near surfaces, and surface instabilities. In all such cases it is important to determine the influence of the boundary conditions.

In finite system (grains for instance), there is no genuine phase transition, since none can take place except in the thermodynamic limit of infinite volume; but there are rounded-off effects of undoubted practical importance, like the so-called collective effects in biological systems, to name only one example. Finite-size effects offer one a way to attack such problems, where fluctuations are very important but fail to induce a fully developed critical regime.

Finally we note that by virtue of displaying crossover behaviour between different spatial dimensionalities, some finite-size effects have an obvious kinship with the effects of spatial anisotropy (Section 7.2).

7.6 Conclusion

In a real system numerous perturbations are generally present. But by virtue of the phenomenon of universality many of these are irrelevant, and by virtue of crossover phenomena it is often possible to isolate the effects due to each relevant perturbation. The chapters which follow will review several different kinds of such perturbations.

References

Essam, J. W. (1972), in *Phase Transitions and Critical Phenomena*, Vol 2, Academic Press.
Fisher, M. E. (1968), *Phys. Rev.*, **176,** 257.
Griffiths, R. B. (1969), *Phys. Rev. Letters*, **23,** 17.
Lubensky, T. C. (1975), *Phys. Rev.*, **B11,** 3219.
McCoy, B. (1972), in *Phase Transitions and Critical Phenomena*, Vol 2, Academic Press.
Wegner, F. J. (1972), *Phys. Rev. B*, **6,** 1891.

CHAPTER 8
Quadratic anisotropy

8.1 Introduction

This chapter deals with systems defined by 'Hamiltonians' of the type

$$\mathcal{H} = \mu_{\dot{0}} + \frac{1}{2} \sum_{i=1}^{n} r_i M_i^2$$

$$+ \sum_{i,j=1}^{n} u_{ij} M_i^2 M_j^2 + \frac{1}{2} \sum_{i=1}^{n} (\nabla M_i)^2 \qquad (8.1)$$

In addition to the quadratic anisotropy in 'spin' space due to the terms involving M_i^2 with $r_i \neq r_j$, we have included anisotropic terms involving $M_i^2 M_j^2$; terms like the latter are in fact generated by the renormalization-group transformations applied to the former. We shall assume that $r_i = r + \Delta r_i$ and $u_{ij} = u + \Delta u_{ij}$, where the Δr_i and Δu_{ij} are functions of an anisotropy field Δ, and vanish when Δ vanishes; thus, in the absence of anisotropy the system reduces to an elementary isotropic system having n components, whose critical behaviour was discussed in Chapter 6.

There are several cases where the Hamiltonian can be written in the form (8.1); they include ferromagnetic systems with anisotropic interactions (Jasnow, Wortis, 1968, Fisher, Pfeuty, 1972), with the Hamiltonian $\mathcal{H} = \sum_{\mathbf{R}} \{ J_X \sigma_{\mathbf{R}}^x \sigma_{\mathbf{R}+\delta}^x + J_Y \sigma_{\mathbf{R}}^y \sigma_{\mathbf{R}+\delta}^y + J_Z \sigma_{\mathbf{R}}^z \sigma_{\mathbf{R}+\delta}^z \}$, and isotropic magnetic systems with an added anisotropic term proportional to $(s^z)^2$; and also Heisenberg antiferromagnets in a uniform field (Fisher, Nelson, 1974).

We consider first the anisotropic Gaussian model with $u_{ij} = 0$ and $r_i = T - T_0 + a_i$, where $a_i > 0$. If one of the a's, say a_λ, is smaller than all the other a_i, then r_λ is the first to vanish; it determines $T_c = T_0 - a_\lambda$, and the order parameter has only one component M_λ. More generally, if there are m equal coefficients r_λ which vanish simultaneously and before the remaining $(n - m)$, then the quadratic anisotropy effectively replaces the isotropic system having n components by a system having m components (Jasnow, Wortis, 1968). We shall show in Section 8.2 that this behaviour remains unaffected by the quartic terms.

8.2 The location of fixed points, and crossover phenomena

Stability of the isotropic fixed point P_n^* with respect to the quadratic anisotropy

We begin by treating the anisotropy as a perturbation, and look at the neighbourhood of the isotropic fixed point of the unperturbed elementary n-component system. According to Wegner's (1972) classification as discussed in

123

Section 7.2, the quadratic anisotropy is a perturbation by a constant field with $m = 0$, $\ell = 2$, $(2m + \ell = 2)$. The anisotropy vector $\{a_i\}$ corresponds to the $(n - 1)$-fold degenerate harmonic polynomial $H_{\ell=2}$; it is orthogonal to the temperature vector $\{1, 1, \ldots, 1\}$, which corresponds to $m = 1$, $\ell = 0$, $(2m + \ell = 2)$. Any arbitrary anisotropy field can then be split into one component which merely shifts the critical temperature, and another orthogonal component which leads to the specific effects we wish to consider. Hence we write the anisotropy operator as $O = \sum_i a_i M_i^2$, where $\sum_i a_i = 0$; the anisotropy field Δ is the conjugate of O.

The fixed point P_n^* is unstable with respect to the temperature, which has dimension $y_{10} = y_E$; it is unstable also with respect to the quadratic anisotropy Δ, which has dimension y_{02}. We saw in Chapter 7 that, taken to first order in ε, y_{02} is positive, is of order of magnitude 2, and is greater than y_{10}; this makes the quadratic anisotropy into a particularly relevant operator. Such strong instability will bring about very pronounced crossover behaviour of the kind described in Section 3.1: as the temperature drops towards $T_c(0)$, the anisotropy field Δ will cease to be negligible at the crossover temperature $\Delta T^* \sim \Delta^{1/\phi}$, the crossover exponent ϕ being given by

$$\phi = \frac{y_{02}}{y_{01}} = 1 + \frac{n\varepsilon}{2(n + 8)} + O(\varepsilon^2) \tag{8.2}$$

Below T^*, as one approaches the critical temperature $T_c(\Delta)$, one observes a new critical behaviour governed by a fixed point more stable than P_n^*. Such new attractive fixed points are well known; they are isotropic fixed points but for a lower spin dimensionality $m < n$. To prove this we must return to the differential renormalization equations in the neighbourhood of $d = 4$.

Differential renormalization equations

To lowest order the differential renormalization equations are

$$\frac{dr_i}{dl} = 2r_i + 16\{2u_{ii}q_i + \sum_j u_{ij}q_i\} \tag{8.3}$$

$$\frac{du_{ij}}{dl} = \varepsilon u_{ij} - 16\{2u_{ij}(u_{ii}q_i^2 + 2u_{ij}q_iq_j + u_{jj}q_j^2) + \sum_k u_{ik}u_{jk}q_k^2\} \tag{8.4}$$

where

$$q_i = \frac{1}{1 + r_i}.$$

For $u_{ij} = u$, and $r_i = r$ small, these reduce to the Equations (5.2) and (5.3) of the isotropic case.

In the presence of anisotropy, however weak, it is no longer possible to choose initial values of the r_i and u_{ij} so as to guarantee convergence to a finitely-distant fixed point. But one can still choose a temperature $T = T_c(\Delta)$ such that, in the subspace associated with a reduced number $m < n$ of components, $(r_i, u_{ij}$ with

$1 \leqslant i \leqslant m$), the critical trajectory converges to a fixed point P_m^*; here m depends on the anisotropy vector $\{a_i\}$, as in the anisotropic Gaussian model. Coordinates r_i, u_{ij} carrying any of the remaining $(n - m)$ indices then tend to infinity. The corresponding factors q_i tend to zero; the differential equations (8.3 and 8.4) decouple and thereby come to describe an isotropic system with m replacing n. Hence the new fixed points are the P_m^* with $m < n$.

Trajectories and crossover

If the anisotropy field Δ is large, the critical trajectory ($T = T_c(\Delta)$) goes directly to the fixed point P_m^*. On the other hand if the anisotropy is weak, the critical trajectory displays crossover behaviour; first it approaches the isotropic fixed point P_n^*, but once arrived in that vicinity it veers away and travels to the fixed point P_m^* which is its ultimate destination, determined by the symmetry of the anisotropy vector $\{a_i\}$. Its sojourn in the vicinity of P_n^* is the longer the smaller Δ and the smaller ϕ. These two extremes of behaviour are illustrated in Figure 8.1.

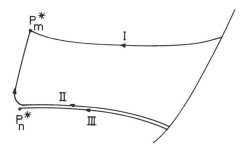

Figure 8.1. Sketch of different kinds of critical trajectories (the trajectories I, II, III correspond, respectively, to Δ large, Δ small and $\Delta = 0$)

Examples: $n = 3$ and $n = 2$

For $n = 3$, one possible source of quadratic anisotropy is a term

$$\Delta(M_1^2 - \tfrac{1}{2}(M_2^2 + M_3^2));$$

in this case, P_2^* is the stable fixed point for positive Δ, and P_1^* for negative Δ. In fact parameter space then contains, in addition to the fixed point P_3^*, three Ising-type fixed points P_1^*, and three P_2^*-type fixed points corresponding to the three different pairings of the axes 1, 2, 3.

For $n = 2$, the quadratic anisotropy term is $\Delta(M_1^2 - M_2^2)$. Then, for either sign of Δ, the stable fixed point P_1^* is of the Ising type. Figure 8.2 shows the critical surface with critical trajectories for $\Delta < 0$.

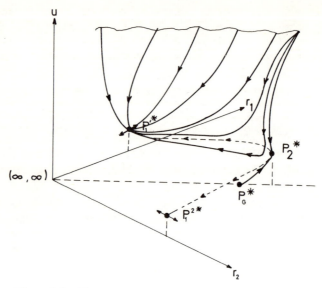

Figure 8.2. The parameter space (r_1, r_2, u), and sketch of the behaviour of the critical trajectories (for $n = 2$ and $\Delta < 0$)

8.3 Homogeneity properties

Because m of the components of the order parameter decouple from the remaining $(n - m)$, one must now distinguish between the parallel susceptibility $\chi_\parallel$ and the transverse susceptibility $\chi_\perp$, which are different functions of T, as shown in Figure 8.3. Above the crossover temperature T^* the anisotropy fails to make itself felt and one has $\chi_\parallel \sim \chi_\perp$. At $T_c(\Delta)$, $\chi_\parallel$ diverges while $\chi_\perp$ remains finite. Applying the arguments of Chapter 4 and Section 3.1 to the differential renormalization equations (8.3) and (8.4) in the neighbourhood of the isotropic fixed point P_n^*, we obtain behaviour that is homogeneous in the variables Δ and $t = \dfrac{T - T_c(0)}{T_c(0)}$; for

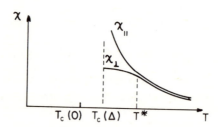

Figure 8.3. Variation of the susceptibilities $\chi_\parallel$ and $\chi_\perp$ as functions of temperature (T^* is the crossover temperature)

instance, $\chi_\parallel$ and $\chi_\perp$ can be written

$$\chi_\parallel(t, \Delta) \sim t^{-\gamma_n} f\left(\frac{\Delta}{t^\phi}\right)$$

$$\chi_\perp(t, \Delta) \sim t^{-\gamma_n} g\left(\frac{\Delta}{t^\phi}\right)$$

(8.5)

Scaling functions

The scaling functions $f(x)$ and $g(x)$ are analytic in the neighbourhood of $x = 0$, and their derivatives behave as follows:

$$\left.\frac{\partial^p \chi_\parallel}{\partial \Delta^p}\right|_{\Delta=0} \sim t^{-\gamma_n - p\phi} f^{(p)}(0)$$

$$\left.\frac{\partial^p \chi_\perp}{\partial \Delta^p}\right|_{\Delta=0} \sim t^{-\gamma_n - p\phi} g^{(p)}(0)$$

(8.6)

One can go further by assuming that the scaling behaviour (8.5) obtains not only locally, i.e. near the fixed point P_n^*, but that it describes also the passage from one fixed point to the other. Near $T_c(\Delta)$, the susceptibility $\chi_\parallel$ diverges like $\dot{t}^{-\gamma_m}$, where $\dot{t} = \frac{T - T_c(\Delta)}{T_c(\Delta)}$, and γ_m is the new exponent; $\chi_\perp$ remains finite. If scaling is global, then this new behaviour implies, on the one hand, that the shift t_Δ the critical temperature varies like

$$\frac{T_c(\Delta) - T_c(0)}{T_c(0)} \sim \bar{w} \Delta^{1/\psi}$$

(8.7)

where the shift exponent ψ is equal to ϕ; and on the other hand, that the scaling function $f(x)$ diverges at $x = \bar{x} = \bar{w}^{-\phi}$ like

$$f(x) \sim (\bar{x} - x)^{-\gamma_m}$$

(8.8)

It follows that $\dot{t}$ and Δ tend to zero, while $\chi_\parallel$ and $\phi_\perp$ vary like

$$\chi_\parallel \sim \Delta^{-(\gamma_n - \gamma_m)/\phi} \dot{t}^{-\gamma_m}, \qquad \chi_\perp \sim \Delta^{-\gamma_n/\phi}$$

(8.9)

Thus $\chi_\parallel$ and $\chi_\perp$ diverge ($\gamma_n > \gamma_m$) like powers, with indices involving the critical exponents γ_n, γ_m and ϕ.

The shift in critical temperature

As the anisotropy field Δ tends towards zero, the critical trajectory labelled II in Figure 8.1 comes extremely close to the fixed point P_n^*, which is a saddle point; after this approach the trajectory travels to the stable fixed point P_m^* practically along the line of steepest descent. In the space spanned by the parameters r_1 and r_2 (with u fixed, $u = u^*$), the equation of the critical trajectory can be written, near P_n^*

(say for $n = 2$), as

$$r_1 + r_2 - 2r^* = C(r_1 - r_2)^{1/\phi} \tag{8.10}$$

Here, the deviations $r_1 - r^*$ and $r_2 - r^*$ from the fixed point are analytic functions of the physical parameters Δ and $t_\Delta = \dfrac{T_c(\Delta) - T_c(0)}{T_c(0)}$. We express this by

$$
\begin{aligned}
r_1 + r_2 - 2r^* &= at_\Delta + b\Delta \\
r_1 - r_2 &= a't_\Delta + b'\Delta
\end{aligned}
\tag{8.11}
$$

where a' vanishes because by symmetry the condition $\Delta = 0$ entails $r_1 = r_2$, whether or not the trajectory is critical. Equations (8.10) and (8.11) yield the shift t_Δ of the critical temperature:

$$t_\Delta \sim a''\Delta + b''\Delta^{1/\phi} \tag{8.12}$$

$\phi = 1$ is a limiting case; for $\phi > 1$ one has $t \sim \Delta^{1/\phi}$, whence $\phi = \psi$, as happens for quadratic anisotropy; for $\phi < 1$ the linear term dominates, whence $\psi = 1 \neq \phi$, as happens for instance for cubic anisotropy as discussed in the next chapter. Then scaling must be reformulated; Equations (8.5) retain their form, but the temperature variable $t = \dfrac{T - T_c(0)}{T_c(0)}$ is replaced by $\dot{t} = \dfrac{T - T_c(\Delta)}{T_c(\Delta)}$.

Experimental tests

It has not proved possible so far to measure experimentally either the crossover exponent ϕ or the scaling functions $f(x)$ and $g(x)$ in anisotropic ferromagnets ($n = 3$), where the anisotropy field Δ is fixed by the nature of the material and hence hard to vary. However there is one case, discussed further in Chapter 12, where one has some hope of measuring ϕ. Anisotropic antiferromagnetics in a uniform external field undergo a first-order 'spin-flop' transition; the attendant discontinuity in magnetization decreases as one approaches the end-point of the first-order transition line, tending to zero with an exponent directly related to ϕ (Fisher, Nelson, 1974). We saw in Chapter 6 that the exponent is accessible also through studying the correlation function when, below T_c, $k\xi$ is large.

By contrast, much information is available from numerical calculations on classical spin systems with $n = 3$ and $n = 2$ on cubic lattices. These showed that the critical exponents have discontinuities at $\Delta = 0$ (Jasnow, Wortis, 1968), and then went on to confirm the scaling properties, to determine the crossover exponent ϕ (giving $\phi = 1.25$ for $n = 3$) and to determine the scaling functions $f(x)$ and $g(x)$ (Pfeuty, Jasnow, Fisher, 1974). Numerical tests of the expressions (8.6) yield estimates of the exponent ϕ and of the derivatives $f^{(p)}(0)$; the latter amount to a Taylor expansion of $f(x)$ which in the light of (8.8) contains information about γ_m and $\bar{x}$. Similar analyses with very weak anisotropy have confirmed the expressions (8.7) and (8.9), and make the equality very plausible.

8.4 Further results

Expansions in powers of ε

The crossover exponent has been calculated to order ε^2 (Wilson, 1972):

$$\phi = 1 + \frac{n}{2(n+8)}\varepsilon + \frac{n(n^2 + 24n + 68)\varepsilon^2}{4(n+8)^3} + O(\varepsilon^3) \qquad (8.13)$$

The exponent ϕ is equal to 1 when $d > 4$; when d < 4 it is greater than 1 but less than γ, tending to γ as $n \to \infty$. If in (8.13) we set $\varepsilon = 1$, (i.e. $d = 3$), and $n = 3$, (as for the Heisenberg model), we find $\phi = 1.22$, in good agreement with the value of $\phi = 1.25$ obtained numerically by Pfeuty, Jasnow and Fisher (1974). Still to be found are the ε-expansions of the scaling functions $f(x)$ and $g(x)$ (Landau theory gives $f(x) = \dfrac{1}{1-x}$).

Anisotropic spherical model; expansion in powers of $1/n$

In the generalized spherical model with the interaction written as $J_\alpha = (1 + g_\alpha)$, $\left(\sum_{\alpha=1}^{l} g_\alpha = 0, l \text{ components } M_\alpha \text{ on each site}\right)$, the crossover exponent ϕ is equal to γ. This is a rather special case, with $\gamma_n = \gamma_m = \gamma = \dfrac{2}{d-2}$, and one cannot in the strict sense of the words speak of competition between two different fixed points. The susceptibility $\chi_{\alpha\alpha} = \dfrac{\partial M_\alpha}{\partial h_\alpha}$ becomes

$$\chi_{\alpha\alpha} = t^{-\gamma}f_\alpha(x_\alpha, x_\beta, \ldots)$$

$$x_\alpha \equiv \frac{g_\alpha}{t^\phi} \qquad (8.14)$$

and the scaling functions obey the system of equations

$$\sum_{\alpha=1}^{l} f_\alpha^{-\gamma} = l, \qquad f_\alpha^{-1} + x_\alpha = f_\beta^{-1} + x_\beta \qquad (8.15)$$

The l-component anisotropic spherical model is equivalent to an $(n \times l)$-component system having l individually isotropic subspaces with an anisotropy to distinguish between them, and taken in the limit $n \to \infty$, l fixed and finite. The exponents and scaling functions can then be expanded in powers of $1/n$. ϕ and ψ have been calculated independently and are indeed equal. To order $1/n$ the expansion can be written

$$\psi = \phi = \frac{4}{3}\gamma - \frac{1}{3}\left(\frac{2}{d-2}\right) \qquad (8.16)$$

where γ is understood to be replaced by its expansion given in Chapter 6.

8.5 Conclusion

In our detailed discussion of the effects due to quadratic anisotropy we have encountered some properties specific to this particular example, and others that are more general and observable in several other cases.

In the presence of quadratic anisotropy, the isotropic fixed point P_n^* is unstable. The crossover exponent ϕ which is a measure of this instability is a new exponent connected with the dimension of the operator $M_i M_j$, $i \neq j$. This exponent, never less than 1 nor greater than γ, corresponds to an $(n-1)$-fold degenerate anisotropy vector. In the immediate vicinity of the critical point the system behaves like an isotropic system of reduced dimensionality $m < n$, corresponding to the stable fixed point P_m^* [recall that there are $(2^n - 1)$ fixed points altogether]. The dimensionality m is determined by the symmetry which survives in presence of the anisotropy. In parameter space the fixed points are infinitely distant from each other. The critical temperature $T_c(\Delta)$ shifts with Δ. When Δ is small, one introduces the shift exponent ψ which in this case is equal to ϕ.

More generally, we have discussed the crossover behaviour and the consequent homogeneity properties, distinguishing the case $\phi > 1$, when $\psi = \phi$, from the case $\phi < 1$, when $\psi = 1 \neq \phi$, leading to more restricted homogeneity properties. This distinction will recur in later chapters.

References

Fisher, M. E., Nelson, D. R. (1974), *Phys. Rev. Letters*, **32,** 1350.
Fisher, M. E., Pfeuty, P. (1972), *Phys. Rev. B*, **6,** 1889.
Jasnow, D., Wortis, M. (1968), *Phys. Rev.*, **176,** 739.
Pfeuty, P., Jasnow, D., Fisher, M. E. (1974), *Phys. Rev. B*, **10,** 2088.
Wegner, F. G. (1972), *Phys. Rev. B*, **6,** 1891.
Wilson, K. G. (1972), *Phys. Rev. Letters*, **28,** 548.

CHAPTER 9
Cubic anisotropy[†]

> 'Amongst the guests were Prometheus, who thinks ahead, and
> Epimetheus, who thinks after the event.'   Physias

9.1 Introduction

In this chapter we consider systems defined by 'Hamiltonians' like

$$\mathcal{H} = \mu_0 + \frac{1}{2}r_0 \sum_i M_i^2 + \frac{1}{2}\sum_i (\nabla M_i)^2 + u\left(\sum_i M_i^2\right)^2 + v\sum_i M_i^4 \qquad (9.1)$$

The coefficient v prefaces a term which is anisotropic in the space spanned by the order parameter; we define this type of anisotropy as 'cubic', from the fact that for $n = 3$ it tends to orient the spontaneous magnetization either along the edges or along the body diagonals of a cube drawn in the space of the order parameter. In this sense the designation 'cubic' can be generalized to arbitrary values of n; to stress that generalization is involved one occasionally speaks of hypercubic anisotropy.

Edges and diagonals

Cubic anisotropy breaks the rotational symmetry in the space spanned by the order parameter. The local free-energy density (i.e. the 'Hamiltonian' $\mathcal{H}$) is no longer invariant under rotations of the field variable at constant magnitude; it assumes stationary values with the field along the edges or the body diagonals, defined respectively by sets of components of the following types:

$$\text{edges:} \quad (1, 0, \ldots, 0)$$

$$\text{body diagonals:} \quad \left(\frac{1}{\sqrt{n}}, \frac{1}{\sqrt{n}}, \ldots, \frac{1}{\sqrt{n}}\right)$$

Along edges the quartic term of the 'Hamiltonian' $\mathcal{H}$ takes the value $(u + v)|\mathbf{M}|^4$, and along diagonals the value $\left(u + \dfrac{v}{n}\right)|\mathbf{M}|^4$. For $\mathcal{H}$ to have a lower bound the two conditions

$$u + v > 0, \qquad u + \frac{v}{n} > 0$$

must be satisfied simultaneously.

[†] Translator's note: The adjective 'cubic' here does not denote 'third order' in the sense in which the adjective 'quadratic' in Chapter 8 denoted 'second order'; the first paragraph of Section 9.1 makes it clear that this accidental contrast in terminology is spurious.

Comparing the values of $(u + v)$ and $\left(u + \dfrac{v}{n}\right)$, we see that diagonals are favoured when v is positive, and edges when v is negative; $(n > 1)$. It is instructive to compare the number of inequivalent edges and body diagonals for different dimensionalities n; inequivalent edges of course number n, while the diagonals number 2^{n-1}, since to define a diagonal we must choose a sign $(\pm)$ for each component and divide by two to avoid double counting. Hence for $n < 2$, diagonals outnumber edges, while they are equal in number for $n = 2$; in the latter case $(n = 2$, but arbitrary dimensionality $d)$, there exists a special symmetry corresponding to an interchange of edges and diagonals.

The case $n = 2$

Let us in this case make a rotation through $\pi/4$ in the space spanned by the order parameter:

$$M'_1 = \frac{M_1 + M_2}{\sqrt{2}} \qquad M'_2 = \frac{M_1 - M_2}{\sqrt{2}}$$

Then the quartic part of $\mathcal{H}$,

$$u(M_1^2 + M_2^2)^2 + v(M_1^4 + M_2^4)$$

expressed in terms of the components M'_1 and M'_2, becomes

$$u'(M_1'^2 + M_2'^2) + v'(M_1'^4 + M_2'^4)$$

where $u' = u + \frac{3}{2}v$ and $v' = v$.

Thus, other things being equal, there is equivalence between two systems specified respectively by the pairs of values (u, v) and $\left(u + \dfrac{3}{2}v, -v\right)$; this equivalence leads to a symmetry which greatly simplifies the investigation of trajectories and fixed points in the renormalization-group approach.

9.2 Fixed points and domains on the critical surface

We shall display our results in the (u, v) plane; in particular we shall consider the projections onto this plane of trajectories and of fixed points located on the critical surface.

We begin by dividing the (u, v) plane into physical and unphysical domains as in Figure 9.1, according to the criterion $\left(u + v > 0 \text{ and } u + \dfrac{v}{n} > 0\right)$ discussed in the last section; note that the second condition involves n.

The two special cases $(v = 0, u > 0)$ and $(u = 0, v > 0)$ correspond to the simple systems which we have discussed already. In the first case we have a simple

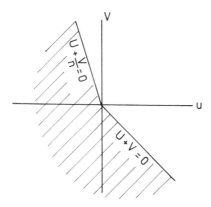

Figure 9.1. Physical and unphysical domains of the (u, v) plane. The unphysical domain is cross-hatched

isotropic system (n, d); in the second case, we have an elementary Ising-type system $(n = 1, d)$, since all the n components are completely decoupled so that the n-fold degeneracy is reflected simply by an overall multiplicative factor in the free energy.

From the results obtained by expansion in powers of $\varepsilon = 4 - d$, we already know three fixed points on the critical surface:

1. The Gaussian fixed point, whose projection onto the (u, v) plane is the origin;

2. the non-trivial isotropic fixed point (n, d) whose projection lies on the axis $v = 0$;

3. the non-trivial Ising-type fixed point $(n = 1, d)$ whose projection lies on the axis $u = 0$.

By using an ε-expansion (Chapter 5) to apply the renormalization-group method to the 'Hamiltonian' (9.1), one discovers a fourth, new, fixed point called the cubic fixed point (Aharony, 1973a).

The location of fixed points

To order ε the four fixed points are located in the (u, v) plane as follows

1. Gaussian fixed point: $\qquad u^* = v^* = 0$

2. Isotropic fixed point: $\qquad u^* = \dfrac{\varepsilon}{16(n + 8)}, \quad v^* = 0$

3. Ising fixed point: $\qquad u^* = 0, \quad v^* = \dfrac{\varepsilon}{16 \times 9}$

4. Cubic fixed point: $\qquad u^* = \dfrac{\varepsilon}{48n}, \quad v^* = \dfrac{\varepsilon}{16 \times 9} \cdot \left(\dfrac{n - 4}{4} \right)$

The result for the cubic fixed point is new: the others have already been given in Section 5.3.

134

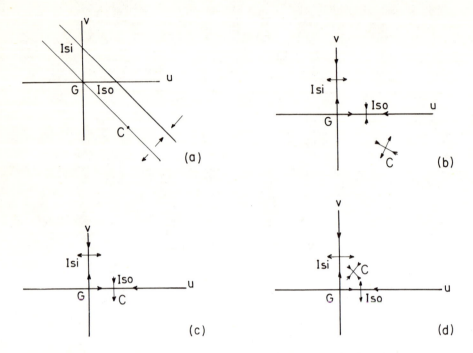

Figure 9.2. Paths of fixed points for varying dimensionality n. (Isi = Ising, Iso = isotropic, G = Gaussian, C = cubic). (a) The case $n = 1$. There is only one variable ($u + v$). The line GC is a ridge, the line Isi–Iso is a valley. (b) The case $n = 2$. G is a peak, Isi and C are saddles, Iso is a hollow. (c) The case $n = n_c(d)$. Iso and C have coalesced, giving rise to a point of inflection for variation along the v-direction. (d) The case $n > n_c(d)$. G is a peak, Isi and Iso are saddles, C is a hollow

It is interesting to follow the positions of these points as n varies, the dimensionality $d = 4 - \varepsilon$ being kept constant; the paths are shown in Figure 9.2.

For $n = 1$, we have only a single variable ($u + v$); the cubic and the Gaussian fixed points are equivalent, and the Ising and isotropic fixed points likewise. For $n = 2$ the special symmetry discussed in Section 9.1 comes into play, and the cubic fixed point is the transform of the Ising fixed point. As n increases, the cubic fixed point moves upwards in the (u, v) plane and coincides with the isotropic fixed point at a particular value n_c of n depending on the dimensionality d:

$$n_c(d) = 4 - 2\varepsilon + \tfrac{5}{2}\varepsilon^2(\zeta(3) - \tfrac{1}{6}) + O(\varepsilon^3) \tag{9.2}$$

To lowest order in ε one has $n_c = 4$ in agreement with the expressions given above.

For $n > n_c(d)$, the cubic fixed point continues upwards in the (u, v) plane and tends to the Ising fixed point as $n \to \infty$. When the cubic and isotropic fixed points coincide they exchange stabilities, exactly as the Gaussian and non-trivial fixed points do, as described in Chapter 5. We shall now discuss this exchange in terms of the values assumed by the anomalous dimensions associated with the various fixed points.

Exchange of stabilities

For each fixed point we indicate, to order ε, the values of the three important anomalous dimensions. These are y_1 (also called y_E); and two others which we denote allusively by y_u and y_v.

1. Gaussian fixed point: $\qquad y_1 = 2, \quad y_u = \varepsilon, \quad y_v = \varepsilon$

2. Isotropic fixed point $\qquad y_1 = 2 - \varepsilon\left(\dfrac{n+2}{n+8}\right), \quad y_u = -\varepsilon$

$$y_v = \varepsilon\left(\frac{n-4}{n+8}\right)$$

3. Ising fixed point: $\qquad y_1 = 2 - \dfrac{\varepsilon}{3}, \quad y_u = \dfrac{\varepsilon}{3}, \quad y_v = -\varepsilon$

4. Cubic fixed point: $\qquad y_1 = 2 - 2\varepsilon\left(\dfrac{n-1}{3n}\right), \quad y_{II} = -\varepsilon,$

$$y_v = \varepsilon\left(\frac{4-n}{3n}\right)$$

Only the values at the cubic fixed point are new; the others have been given already in Sections 5.3 and 7.2. In general the fields u and v are not strictly speaking scaling fields; the notation y_u and y_v is defined by continuity in the limit of small ε. In the notation of Chapter 5 one has $y_{II} = y_u$ for the isotropic fixed point, and $y_{II} = y_v$ for the Ising fixed point.

From the values given above one can determine the stabilities of the fixed points as projected onto the (u, v) plane. The Gaussian fixed point, doubly unstable, is a peak; the Ising fixed point, singly unstable, is a saddle point. For $n < n_c(d)$, the cubic fixed point, singly unstable, is a saddle point, while the isotropic fixed point is stable (at the bottom of a hollow). For $n = n_c(d)$, the latter two coalesce, which is signalled by marginal behaviour of the anisotropy field v. For $n > n_c(d)$, their stabilities have been exchanged and it is the cubic fixed point, now in the upper half-plane, which is stable.

The curve L_v defined by $n = n_c(d)$ is found in practice by determining the values of n for which the anomalous dimension y_v, associated with the fixed point (n, d), vanishes; it has been calculated to order ε^3 (Ketley and Wallace, 1973), whence the formula (9.2). The general appearance of L_v in the (n, d) plane is shown in Figure 9.3.

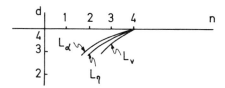

Figure 9.3. Sketch of the curve L_v defined by $n = n_c(d)$. For comparison, the curves L_α and L_η defined in Section 6.2 are also shown

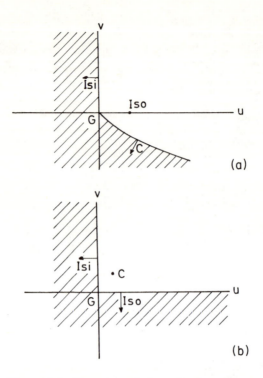

Figure 9.4. Domains of the critical surface projected onto the plane (u, v). (a) The case $n < n_c(d)$. The cross-hatched domain corresponds to first-order transitions. The other domain is the catchment area of the isotropic fixed point, and corresponds to second-order transitions whose critical regime is governed by that fixed point. (b) The case $n > n_c(d)$. The cross-hatched region corresponds to first-order transitions. The other region is the catchment area of the cubic fixed point C, and corresponds to second-order transitions whose critical regime is governed by that fixed point

As regards the physics, it is useful to note that the cubic fixed point is stable when it lies in the upper half-plane, or in other words when it corresponds to a system where the diagonals are favoured. In view of the stabilities of the various fixed points one can divide the (u, v) plane into domains and indicate the critical behaviour in each, as in Figure 9.4. The usual cases $(u > 0, v$ small) are summarized thus:

When $n < n_c(d)$, cubic anisotropy is irrelevant and the critical exponents are those of the isotropic system.

When $n > n_c(d)$, we must distinguish between two possibilities. If the aniso-tropy favours the diagonals, one expects crossover behaviour followed by an ultimate critical regime at the cubic fixed point; but if the edges are favoured, then one expects a first-order transition. Moreover, if by varying some physical parameter one can change the sign of v, then one will observe something like a tricritical point, at which the order of the transition changes, but which is characterized by the critical exponents of the isotropic case (n, d) rather than by the standard tricritical exponents.

The critical exponents governed by the cubic fixed point have been calculated to order ε^2:

$$v_c = \frac{1}{2} + \left(\frac{n-1}{6n}\right) + \frac{(n-1)(17n^2 + 290n - 424)}{648n^3} \cdot \varepsilon^2 + O(\varepsilon^3)$$

$$\eta_c = \frac{(n+2)(n-1)}{54n^2} \varepsilon^2 + O(\varepsilon^3) \tag{9.3}$$

For $n = 1$ these are consistent with the Gaussian exponents, and for $n = 2$ with the Ising exponents, as they must be.

Further, the anomalous dimension y_v has been calculated to order $1/n$:

$$y_v = \varepsilon - \frac{4\varepsilon}{n} \frac{(2-\varepsilon)(6-\varepsilon)}{4-\varepsilon} \cdot A(\varepsilon) + O\left(\frac{1}{n^2}\right)$$

with the function $A(\varepsilon)$ defined in Section 6.2. This expression shows that for large n the field v becomes more relevant as the dimensionality d decreases.

The proximity of the cubic to the Ising fixed point in the limit $n \to \infty$ is reminiscent of the proximity of the non-trivial to the Gaussian fixed point in the limit $\varepsilon \to 0$ (see Chapter 5). One can develop a renormalization-group approach closely following this analogy, and can thus derive the critical exponents in the cubic case by a perturbation method starting in zero order from the Ising system (Aharony, 1973b).

9.3 Comments on the results

In this section we comment on the results obtained by the ε- and $\frac{1}{n}$-expansions, and end by considering the curve L_v and the case $(n = 2, d = 2)$.

Expansions in powers of ε and $1/n$

Consider the exponents v_c and η_c as given by (9.3). Notice that for $n = n_c(d)$, the values of the 'cubic' and of the 'isotropic' exponents coincide; this identity stems from the coincidence of the two fixed points for this value of n. As n varies, one has $\eta_c < \eta_{iso}$ for $n < n_c(d)$, and $\eta_c > \eta_{iso}$ for $n > n_c(d)$. In other words, to this order in ε the more stable fixed point is the one with the greater value of η.

One notes that in the limit $n \to \infty$, η_c unlike η_{iso} fails to vanish. In fact in this limit an exact solution exists for any dimensionality d, with critical exponents

$$\eta_c = \eta_1, \qquad v_c = \frac{v_1}{1 - \alpha_1}$$

where η_1, v_1, α_1 are the critical exponents of the Ising model ($n = 1$). In other words, when $n \to \infty$ the 'cubic' critical exponents are the same as the Ising model exponents after Fisher-type renormalization (see Section 7.3), just as the isotropic exponents arise from such a renormalization of the Gaussian ones. Indeed, from Figure 9.2 one can see that as $n \to \infty$, the cubic approaches the Ising fixed point in the same way as the isotropic fixed point approaches the Gaussian. But these approaches do not lead to coalescence; by erecting a vertical r_0-axis one can see that in this limit the fixed points instead of coinciding lie one above the other. Fisher's renormalization is linked with this kind of superposition, the temperature axes for the two superposed fixed points being orthogonal.

The curve L_v and the case ($n = 2, d = 2$)

How should the above results be extrapolated in the (n, d) table?

First we must clarify our ideas about the curve L_v; so far we know one point on it, plus its tangent and curvature at that point (see Figure 9.3). In view of the special symmetry for $n = 2$, it seems that the curve cannot simply intersect the vertical through $n = 2$; indeed for $n = 2$ the isotropic and cubic fixed points can never coincide because they are confined to different axes in the (u, v) plane. Therefore they could exchange stabilities only in an exotic manner, that is to say through the appearance of a whole line of fixed points joining them. It is doubtful whether this could happen for dimensionalities $d > 2$. Moreover, we can rule out the possibility that as d decreases the curve L_v retreats in the direction of increasing n, because of certain results showing that cubic anisotropy becomes more and more relevant as $n \to \infty$.

At this stage one should mention that Baxter's model (Baxter, 1971) can be pictured as two Ising models coupled by a so-called energy–energy interaction (Kadanoff, Wegner, 1971). It can then be regarded as a model with ($n = 2, d = 2$) having cubic anisotropy, and possesses a marginal operator, with persistent marginality (see Chapter 13); consequently in its parameter space there exists a line of fixed points. In view of how the problem is framed, it is not clear whether the Baxter model contains the isotropic case as one of its special cases. Nevertheless it seems plausible that the curve $n_c(d)$ passes through the point ($n = 2, d = 2$).

This assumption if accepted allows one to deduce the value of the exponent η for ($n = 2, d = 2$). From the fact that cubic anisotropy is a marginal perturbation, one concludes that at T_c, $\eta = \frac{1}{4}$. One knows moreover that $\eta = \frac{1}{4}$ also for the Ising model ($n = 1, d = 2$). Now, in the (n, d) table, the curve L_η is the locus of points where η has maxima as a function of n with d fixed: hence L_η must lie between the two points ($n = 2, d = 2$) and ($n = 1, d = 2$). This is certainly consistent with the results from the ε-expansion, since according to these L_η lies between the curves L_α and L_v.

It will have been noticed that this discussion turns on several rather unusual plausibility arguments. Future work will show whether its predictions are correct. Indeed, our aim in presenting the argument was to illustrate, convincingly we hope, that the topological approach of the renormalization group lends itself not only to rigorous but also to heuristic procedures.

9.4 Conclusion

Cubic anisotropy affects many real phenomena; magnetic and structural phase transitions are examples. Thus the physical importance of such perturbations is unquestioned, and the predictions presented in this chapter can be checked experimentally. From the point of view of renormalization-group methods, cubic anisotropy provides a most telling illustration, especially of the wealth inherent in the topological description.

References

Aharony, A. (1973a), *Phys. Rev.*, *B* **8**, 4270.
Aharony, A. (1973b), *Phys. Rev. Letters*, **31**, 1494.
Baxter, R. J. (1971), *Phys. Rev. Letters*, **26**, 832.
Kadanoff, L. P., Wegner, F. J. (1971), *Phys. Rev.*, *B* **4**, 3989.
Ketley, I. J., Wallace, D. J. (1973), *J. Phys. A*, **6**, 1667.

CHAPTER 10
Perturbations by constant fields

This chapter reviewing perturbations by constant fields leans directly on the introduction given in Section 7.2, and on the two preceding chapters which dealt in detail with quadratic and cubic perturbations. The first section abstracts from the examples already studied a systematic approach to the effects of any arbitrary field. The three subsequent sections are devoted to other important special cases of perturbations by constant fields. Finally, Section 10.5 touches on the vast and somewhat isolated subject of finite-size effects.

10.1 Methodology

Consider a perturbation which contributes to the local free-energy density a term of the form

$$\mu_i \cdot O_i(\mathbf{x})$$

This is the simplest possible case, allowing an unambiguous definition of the field μ_i and the operator O_i. In certain other cases it may be difficult to identify a local density responsible for the perturbation (as for instance for finite-size effects), and hence the definition of the field μ_i may be somewhat arbitrary.

The first step is to determine the effect on the unperturbed system of a weak field μ_i, or more precisely, to determine the relevance of the field μ_i with respect to the unperturbed fixed points. In many cases the crossover exponents ϕ_i can be found solely from the symmetry of the operator O_i.

Having studied the situation in the neighbourhoods of the unperturbed fixed points, one must explore the rest of parameter space. If the field μ_i is relevant, one needs to discover whether the trajectories lead to new fixed points, and if so, to study these. If the field μ_i is irrelevant, one needs to determine the catchment areas and to find out what happens beyond them. In this way one discovers whether the perturbation can modify the critical behaviour or change the order of the transition. If the field μ_i is marginal, one must see whether the marginality is local or persistent, and draw the appropriate conclusions, which can be very varied (see Chapter 13).

Next one must formulate the homogeneity rules, taking into account any crossover effects due to the field μ_i. This programme includes the calculation of the exponent ψ for the shift in T_c, and the comparison between ϕ and ψ, which determines the kind of scaling appropriate to the problem.

Finally one must examine the experimental evidence, obtained either in the laboratory or by numerical calculation.

The examples we have discussed, and those to follow, demonstrate that very often there is no need to start from scratch. For quadratic anisotropy (Chapter 8), the crossover exponent was new but the fixed points were of an already known kind. For cubic anisotropy (Chapter 9), a new kind of fixed point did appear, but general considerations allowed its novel features to be circumscribed. For spatial anisotropy (Section 7.2), the crossover exponent and the fixed points were already known. While the concepts of the renormalization group provide a very useful theoretical framework for correlating a wide variety of arguments, the role of successful calculations and of ε-expansions is often simply supportive or confirmatory.

10.2 Long-range forces

We have seen that in the presence of long-range forces of the kind described in Section 2.5, namely

$$J(r) \sim \frac{1}{r^{d+\sigma}}$$

the contribution to the free energy from fluctuations M_k having momentum k contains a term

$$\mu_\sigma . k^\sigma |M_k|^2, \quad \text{provided that } \sigma < 2.$$

The field associated with the perturbation due to such forces is then, by definition, the coefficient μ_σ. Note that this generalizes the concept of a field, in that the perturbation, because of its long-range character, cannot be expressed as the product of a field and a local operator.

We can then find the anomalous dimension y_σ of the field μ_σ with respect to the short-range fixed point in terms of the anomalous dimension of M_k:

$$y_\sigma = 2 - \eta - \sigma$$

This determines the crossover exponent

$$\phi_\sigma = y_\sigma . v = (2 - \eta - \sigma)v$$

When $\sigma = 2 - \eta$, the field μ_σ is marginal, reflecting the coincidence of the short-range fixed point with a new, namely with the long-range, fixed point. When $\sigma < 2 - \eta$, it is the long-range fixed point which is stable.

For $\sigma < d/2$, there is another exchange of stabilities between two long-range fixed points, the trivial and the non-trivial (see Section 2.5). Figure 10.1 shows the regions of the (d, σ) plane which correspond to the various critical regimes. It remains only to elucidate the critical regime in region IV, i.e. the regime governed by the non-trivial long-range fixed point. It seems that the exponent η_{LR} is given by

$$\eta_{LR} = 2 - \sigma$$

i.e. that it retains its classical long-range value which is independent of n and d; in any case this result is consistent with the expansions in powers of $\varepsilon = 2\sigma - d$ and of

142

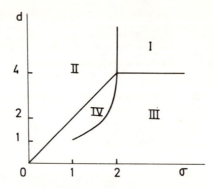

Figure 10.1. Regions of the plane (d, σ) corresponding to the various critical regimes. Regions I, II, III, IV correspond, respectively, to the classical short-range regime, the classical long-range regime, the non-trivial short-range regime and the non-trivial long-range regime. The exact position of the boundary between regions III and IV depends on the dimensionality n of the order parameter

$1/n$. As regards the exponent γ_{LR}, the expansion in $\varepsilon = 2\sigma - d$ gives

$$\frac{1}{\gamma_{LR}} = 1 - \left(\frac{n + 2}{n + 8}\right) \cdot \frac{\varepsilon}{\sigma} - \frac{(n + 2)(7n + 20)}{(n + 8)^3} A(\sigma) \cdot \left(\frac{\varepsilon}{\sigma}\right)^2 + O(\varepsilon^3)$$

where $A(\sigma) = \sigma\left[\psi(1) - 2\psi\left(\frac{\sigma}{2}\right) + \psi(\sigma)\right]$, with $\psi(x)$ the digamma function (Fisher,

Ma, Nickel, 1972). For $n \to \infty$ one has

$$\gamma_{LR} = \frac{\sigma}{d - \sigma}$$

Suzuki (1973) has calculated also the term proportional to $1/n$.

As usual, the boundaries in Figure 10.1 correspond to marginal situations where logarithmic corrections can occur.

The results for the one-dimensional Ising model ($d = 1$) are in full harmony with this general discussion. In that case it has been shown directly (Dyson, 1969) that there exists a classical regime for $\sigma < \frac{1}{2}$, a short-range regime for $\sigma > 1$, (consistently with the value $\eta = 1$), and an intermediate regime for $\frac{1}{2} < \sigma < 1$. Note that when $\sigma > 1$ there is no phase transition at finite temperature, whence the value $\sigma = 1$ (i.e. interaction $\sim 1/r^2$) divides one regime where $T_c \equiv 0$ from another where $T_c \neq 0$. It seems that for this special value itself there is a discontinuity in the order parameter at T_c, (see the discussion of the Kondo effect in Section 13.3).

10.3 Dipole–dipole forces

Dipole–dipole forces are long-range, but differ from the long-range forces considered so far through not being invariant under independent rotations in ordinary space and in the space spanned by the order parameter. The alternations in sign of dipole forces reduce their effects compared with those due to perturbations of constant sign.

From the outset we distinguish between the case of ferromagnets and of antiferromagnets. For an antiferromagnetic phase transition one finds the fairly obvious result that dipole forces are irrelevant, precisely because of cancellations between contributions of opposite sign. But in ferromagnetic transitions the dipole forces introduce a demagnetizing field, or in other words important cumulative effects. We deal separately with the isotropic case ($n = d$) and the uniaxial case ($n = 1$, d arbitrary).

The isotropic case (Aharony, Fisher, 1973) must have $n = d$ because of the coupling between real space and the space spanned by the order parameter. The demagnetizing field due to the dipole forces adds to the external field H a correction term NM; the coefficient of demagnetization, N, is proportional to the strength of the dipole–dipole forces and has the dimensions reciprocal to susceptibility. It follows that the crossover exponent for the dipolar perturbation is

$$\phi = \gamma$$

showing that the perturbation is highly relevant. Accordingly there appears a new, the dipolar, fixed point, whose critical exponents have been calculated in an ε-expansion (recall that the condition $n = d$ is being maintained):

$$\eta_{\mathrm{D}} = \frac{20}{867} \cdot \varepsilon^2 + O(\varepsilon^3)$$

$$\gamma_{\mathrm{D}} = 1 + \frac{n + 2}{2\left[n + 8 - \dfrac{4}{n + 2} \right]} \varepsilon + O(\varepsilon^2)$$

Thus one reaches the unexpected conclusion that dipole forces increase rather than decrease the deviations of the exponents from their classical values, contrary to what might have been inferred from the long-range nature of such forces.

In the uniaxial case, dipole forces have a more drastic effect, in that they change the characteristic dimensionality to $d_{\mathrm{c}} = 3$, (Larkin, Khmelnitskii, 1969). For $d = 3$ one finds logarithmic corrections to the classical expressions, and for $d < 3$, new exponents which can be calculated through expansions in powers of $(3 - d)$.

10.4 The Dzyaloshinskii–Moriya interaction

The coupling in question is of the type $\mu_{\mathrm{DM}}\mathbf{A} \cdot (\mathbf{S}_i \times \mathbf{S}_j)$, and has been used to account for certain complex magnetic structures. Once this vectorial coupling between components of the order parameter is given, it becomes convenient to impose the condition $n = 3$ from the outset. The relevance of the field μ_{DM} with respect to the unperturbed fixed point can be guessed on the usual dimensional grounds; for the crossover exponent one finds

$$\phi = \phi_{QA} - \nu$$

where ϕ_{QA} is the crossover exponent for quadratic anisotropy (see Chapter 8). Hence in general the field μ_{DM} is relevant. The fixed point which is stable in presence of this perturbation is of a kind we already know, being like the isotropic fixed point ($n = 2$); it leads to helical ordering, where the order parameter is a density wave possessing amplitude and phase in fairly close analogy to the order parameter of smectic A phases.

If the initial problem contains quadratic anisotropy as well (Liu, 1973), then, depending on the strengths of the different fields, different kinds of order can ensue corresponding to $n = 1$, $n = 2$ or $n = 3$. Qualitatively the situation is the same as if the quadratic anisotropy alone were present, with the field μ_{DM} affecting the position of the watershed between the catchment areas of the fixed points $n = 1$ and $n = 2$.

10.5 Finite-size and surface effects

We shall concentrate particularly on the properties of films that are infinitely extended in $(d - 1)$ dimensions and are finite in one dimension. Many of the results generalize readily to other geometries.

The presence of a finite dimension raises the problem of what boundary condition to choose, and whether the results are universal, i.e. independent of this choice. In practice we shall consider mainly boundary conditions that are either periodic, or correspond to free surfaces with or without modification of the interactions in the surface region.

It is convenient to distinguish four different categories of finite-size effects. The first category concerns modifications of bulk properties due to the finite extent in one dimension, namely shifts of T_c and of the crossover temperature. The second concerns surface contributions which add to the bulk contributions to the thermodynamic variables. The third concerns the properties of the surface itself. The fourth category concerns surface instabilities as distinct from volume instabilities; these can arise for certain kinds of boundary conditions. Notice that effects in the last three categories are not peculiar to finite size, since they persist for instance in semi-infinite media.

Evidently we are faced with a vast subject which offers scope to the whole arsenal of methods developed for infinite media in the preceding chapters: Landau theory, phenomenological homogeneity rules, scaling laws, exactly soluble models, numerical work, and suitably adapted renormalization-group methods.

Shifts in T_c and crossover

Consider a film infinite in $(d - 1)$ dimensions and of finite length L in one dimension. As long as the correlation length ξ is less than L, the film behaves globally like a medium of dimensionality d. The crossover to the true dimensionality $(d - 1)$ occurs when

$$\xi \sim L$$

allowing us to define a crossover temperature by

$$(\Delta T)^* \sim \left(\frac{1}{L}\right)^\theta$$

Here, $\theta = 1/v$, the exponent v being the critical exponent for dimensionality d.

It is convenient to consider θ as a crossover exponent of the usual kind, which implies that $\left(\frac{1}{L}\right)$ is the perturbing field. Actually it might be advantageous to define the finite-size field differently, for instance as some power of $\frac{1}{L}$, but we shall not pursue this question. Our only precaution will be to write θ and λ instead of the usual symbols ϕ and ψ.

The argument determining the exponent θ is clearly independent of the boundary conditions. The situation is less simple as regards the shift in T_c. Here one defines an exponent λ by

$$T_c(0) - T_c\left(\frac{1}{L}\right) \sim \left(\frac{1}{L}\right)^\lambda$$

The opinion prevalent at present, in view of various special results, is that for boundary conditions corresponding to a free surface one has

$$\lambda = \theta = \frac{1}{v}$$

while for periodic boundary conditions (Domb, 1973),

$$\lambda = d - 2 + \eta$$

Periodic boundary conditions are less perturbing than those for a free surface, and lead to a smaller shift in T_c.

Note that for ideal Bose condensation and for the spherical model, with free surfaces, one finds

$$\lambda = 1$$

This result, hinging on the global nature of the constraints on the models, would seem to suggest that the analogy with the limit $n \to \infty$ fails to extend to certain finite-size effects. Besides, under such conditions one obtains for these models the bizarre result that T_c increases instead of decreasing as one would have expected.

There is one situation deserving mention, where the $(d - 1)$-dimensional system either has no transition or has a transition of a special kind (e.g. for $d = 3$ and $n > 1$, short long-range order). We note only that the ideas introduced above can be extended to such systems without too much difficulty.

Surface contributions

We expect that the free energy for a film of thickness L can in the limit of large L be written as

$$G = G_\infty + \frac{2}{L}\tilde{G} + \cdots$$

where G_∞ and $\tilde{G}$ are the bulk and the surface contributions, respectively; and similarly for the other thermodynamic variables. Accordingly one defines new exponents $\tilde{\alpha}$, $\tilde{\beta}$, $\tilde{\gamma}$ associated with the singular behaviour of the surface contributions.

Scaling laws can easily be devised by dimensional analysis; thus,

$$\tilde{G} \sim \left(\frac{1}{\xi}\right)^{d-1} \sim (\Delta T)^{(d-1)\nu}$$

leads to $\tilde{\alpha} = \alpha + \nu$; similarly $\tilde{M} \sim M\xi$ and $\tilde{\chi} \sim \chi\xi$ lead to

$$\tilde{\beta} = \beta - \nu, \qquad \tilde{\gamma} \sim \gamma + \nu$$

In general these predictions appear to be confirmed; once again the Bose and the spherical models play a distinct role, with ν replaced by 1 in the three expressions just above. Exactly as for infinite media, these scaling laws can be derived from phenomenological homogeneity rules which we shall discuss in the next subsection.

Finally some comments. For periodic boundary conditions the surface contributions defined as above are identically zero. For ordinary fluids the scaling laws predict the vanishing of the surface tension at the liquid-to-gas critical point. In systems with $n > 1$ there are some peculiarities stemming from directional (angle) variables; thus, in zero field below T_c, the surface contribution to the spontaneous magnetization diverges, which is connected with the divergence of the parallel susceptibility (Section 6.3) and with the infinite range (power-law behaviour) of the perturbations due to the surface.

Properties of the surface

Properties of the surface are interesting in themselves. One example is the magnetization of the surface atoms, as distinct from the change in bulk magnetization studied in the last subsection. Such local properties are important because they are measurable experimentally, and because they can influence physical and chemical processes occurring near the surface.

In terms of the surface magnetization M_1 (i.e. the magnetization of a surface atom) one defines an exponent β_1 by $M_1 \sim (\Delta T)^\beta$; in terms of the susceptibility χ_1 (the response of M_1 to a uniform field H), one defines γ_1 by $\chi_1 \sim (\Delta T)^{-\gamma_1}$; in terms of the susceptibility $\chi_{1,1}$, which is the response of M_1 to a field H_1 coupled only to the surface atoms, one defines $\gamma_{1,1}$ by $\chi_{1,1} \sim (\Delta T)^{-\gamma_{1,1}}$; the exponent δ_1 by $H_1 \sim M_1^{\delta_1}$ at T_c; the gap exponent Δ_1 by $\Delta_1 = \beta_1 \delta_1$; and finally exponents $\eta_\parallel$ and

$\eta_\perp$ in terms of the rates of decrease of the correlation function parallel and perpendicular to the surface.

It must be said at once that these exponents apparently cannot be expressed in terms only of the bulk exponents, differing in this respect from exponents like $\tilde{\alpha}$ in the last subsection. Nevertheless, from the phenomenological homogeneity rules one can derive scaling laws interconnecting the new exponents, so that knowledge of any one of them, plus knowledge of the bulk exponents, seems sufficient to determine all the others.

The singular part of the free energy is given by the expression

$$G(T, H, L, H_1) = (\Delta T)^{2-\alpha} \cdot g\left(\frac{H}{(\Delta T)^\Delta}, L \cdot (\Delta T)^\nu, \frac{H_1}{(\Delta T)^{\Delta_1}}\right)$$

which is perfectly in keeping with the ideas of Chapter 3. By setting $H = 0$ in such a homogeneity rule we can derive scaling laws for surface contributions, discussed in the last subsection. In the presence of H_1 we obtain additional scaling laws for the new exponents, like

$$\beta_1 + \gamma_1 = \Delta; \beta_1 + \gamma_{1,1} = \Delta_1; \beta_1 = 2 - \alpha - \nu - \Delta_1; \gamma_1 = (2 - \eta_\perp)\nu$$

$$\gamma_{1,1} = (1 - \eta_\parallel)\nu$$

At this stage it is tempting to apply the programme of Chapter 3 by expressing the homogeneity rules as invariance properties under dilatation. But the fact that translational invariance is now broken introduces the complication that the correlation function depends on the absolute positions of two points, and not merely on their relative position. In the renormalization-group approach this complication is basic. One can see this by recalling the steps described in Chapter 4, and attempting to adapt them to a semi-infinite and therefore inhomogeneous system. Such adaptation is in fact possible (Lubensky, Rubin, 1973) and for an Ising system ($n = 1$) one finds to first order in $\varepsilon = 4 - d$

$$\eta_\perp = 1 - \frac{\varepsilon}{\sigma} + O(\varepsilon^2)$$

Recall the classical values of the exponents for an ordinary critical point:

$$\beta_1 = 1, \gamma_1 = \tfrac{1}{2}, \gamma_{1,1} = -\tfrac{1}{2}, \Delta_1 = \tfrac{1}{2}, \eta_\parallel = 2, \eta_\perp = 1$$

A review of soluble models and of numerical calculations has been given by Fisher (1973).

Surface instabilities

In the preceding subsections we have been interested in critical behaviour near transitions of the system as a whole. But for certain boundary conditions it can happen that besides such global transitions at T_c there appears also a surface transition at a temperature T_s different from T_c. Perhaps the best studied example is surface superconductivity, but one should mention also the phenomena of reconstruction and superstructure in crystals, magnetic instabilities, etc., currently under active study (Blandin, 1973). The order parameter for such surface transitions decreases as one moves away from the surface towards the interior. The effective spatial dimensionality is $(d - 1)$, i.e. the dimensionality of the surface; hence one expects the results corresponding to infinite media but with dimensionality $(d - 1)$. Variation of the boundary condition produces crossover phenomena when T_s approaches to T_c.

Experimental aspects

To conclude this section we must note that there is much topical interest in such surface effects, arising from the considerable recent progress in surface physics generally. Improvements in techniques for preparing and studying surfaces are leading to many new measurements. Moreover, since critical fluctuations become increasingly significant in spaces of lower dimensionality, it is very possible that an important new field of study is emerging in the region where the theories of critical phenomena and of heterogeneous catalysis overlap (Suhl, 1973).

10.6 Conclusion

It is difficult to conclude a chapter that has dealt with so varied a range of topics, except by stressing that the effects predicted theoretically call for experimental tests, and that in some cases, especially those discussed in Section 10.5, the effects could be of great practical importance.

References

Aharony, A., Fisher, M. E. (1973), *Phys. Rev., B8*, 3323.
Blandin, A. (1973), in *Collective Properties of Physical Systems*, Nobel Symposium XXIV, Academic Press.
Domb, C. (1973). *J. Phys. A*, **6,** 1296.
Dyson, F. J. (1969), *Comm. Math. Phys.,* **12,** 91.
Fisher, M. E., Ma, S. K., Nickel, B. G. (1972), *Phys. Rev. Letters,* **29,** 917.
Fisher, M. E. (1973), *J. Vac. Sci. Tech.,* **10,** 665.
Larkin, A. I., Khemelnitskii, D. E. (1969), *Sov. Phys. J.E.T.P.,* **29,** 1123.
Liu, L. L. (1973), *Phys. Rev. Letters,* **31,** 459.
Lubensky, T. C., Rubin, M. H. (1973), *Phys. Rev. Letters,* **31,** 1469.
Suhl, H. (1973), in *Collective Properties of Physical Systems*, Nobel Symposium XXIV, Academic Press.
Suzuki, M. (1973), *Progr. Theor. Phys.*, **49,** 1106.

CHAPTER 11
Coupling to other degrees of freedom

A magnetic transition, to take one example, is often accompanied by a change in the lattice parameter. One says then that the magnetic degrees of freedom are coupled to the translational degrees of freedom of the atoms, and that because of this coupling the atomic displacements depend on the magnetic order. According to this formulation some degrees of freedom are dominant (in this case the magnetic), while others play only a secondary role (here the elastic degrees of freedom). The title of the present chapter reflects the existence of such a hierarchy. In many cases it is legitimate to start by considering only the dominant degrees of freedom, and only subsequently to examine the perturbation due to their couplings to others.

However, there are also more democratic situations where the several degrees of freedom play equally important roles. For instance, the problems of cubic anisotropy discussed in Chapter 9 can be ascribed to a coupling between different degrees of freedom, namely different components of the order parameter, which enter on an equal footing.

In the case of tricritical points, studied in the next chapter, it usually seems possible to distinguish two different kinds of mutually interacting order parameters. Thus for the tricritical points of (^{3}He, ^{4}He) mixtures there is one superfluid order parameter (embodying two degrees of freedom), and another governing phase-separation (one degree of freedom). This analysis generalizes to polycritical points.

Finally there are many systems where fluctuations of different kinds are important simultaneously, and enter into competition. In metals, for instance, the Peierls, superconducting, and antiferromagnetic transitions all tend to produce a forbidden band at the Fermi level, resulting in an interaction between these different degrees of freedom which is especially important in one-dimensional metals (see Section 13.4). It is still too early to see to what extent renormalization-group methods will be able to predict from first principles the various possible types of order, and their domains of existence, in such very complex systems (Anderson, 1973).

Having stressed how extensive is the range of problems involving coupled degrees of freedom, we shall confine ourselves in the following sections to some rather simple ideas, (like Landau's method of eliminating the secondary degrees of freedom), and to some instructive special results.

11.1 Spin–phonon (magnetoelastic) couplings

Several different kinds of coupling are possible, and different models lead to different results. It is conceivable that there is more universality than is apparent, and that some models must simply be discarded as too simplified to represent real couplings. To be more precise, it is possible that all the different kinds of couplings must be incorporated simultaneously and that there is one amongst them which in general will be dominant. Be that as it may, we shall be led to distinguish between isotropic and anisotropic, and between local and long-range couplings.

In this discussion we attribute the dominant role to the magnetic degrees of freedom; the elastic degrees of freedom will be eliminated through changes in the magnetic coupling constants, which may entail in turn changes in the order of the transition or in the critical behaviour. But we should mention that in certain other models of magneto-elastic couplings it proves more convenient to eliminate the magnetic and retain the elastic degrees of freedom; this is so especially in some one-dimensional models (Pincus, 1971).

Landau method. Elimination of the elastic variables

Consider the following local free-energy density

$$F = F_0 + \tfrac{1}{2}r_0 M^2 + u_0 M^4 + \tfrac{1}{2}E.\theta^2 + A\theta M^2 \tag{11.1}$$

Gradient terms and the distinction between different components of the magnetization field-variable M have been neglected, since we intend in any case to use the Landau method, as in Section 2.3. The elastic variable θ is the local dilatation of the lattice, coupled to the local magnetic energy through the coupling constant A.

At constant M the free energy F is minimized by

$$\theta = -\frac{A}{E}.M^2$$

Substituting this optimal value of θ into (11.1) we find

$$F = F_0 + \tfrac{1}{2}rM^2 + \tilde{u}_0 M^4$$

where the effective parameter $\tilde{u}_0$ is given by

$$\tilde{u}_0 = u_0 - \tfrac{1}{2}\frac{A^2}{E}$$

Thus the effects of the elastic variables are represented by a decrease in the local magnetic coupling constant u_0. When the coupling A is weak, the order of the transition remains unchanged; at a certain value of A a tricritical point appears; for greater values of A the transition becomes first order.

These results for local and isotropic magnetoelastic couplings are confirmed qualitatively by an analysis which also takes account of fluctuation effects.

Baker–Essam models

These are compressible Ising models with boundary conditions that can be adjusted to represent either constant-volume or constant-pressure constraints, and they are available in different versions; (Baker, Essam, 1970, 1971; Gunther *et al.*, 1971). The couplings take effect essentially through changes in the overall elastic properties, whence the boundary conditions at the surface are important; this global character contrasts with the local character of the couplings discussed in the preceding subsection. In fact the properties of a compressible Ising structure follow from those of the rigid Ising structure, with the boundary conditions assuming the role of global constraints in the sense of Section 7.3. Thus in the original Baker–Essam model one observes a Fisher-type renormalization under conditions of constant positive pressure or of constant volume; under zero pressure there is no renormalization; and under negative pressure the transition is of first order. The transition at zero pressure looks like a special kind of tricritical point. But all these results should be treated with caution in view of the simplifications which are built into the models from the outset in order to make them soluble.

Magnetothermomechanics

This portly designation covers a range of investigations into thermomechanical instabilities (Wagner, Swift, 1970). If one expresses realistically the interactions responsible for the kinds of global effects described in the last subsection, one encounters two types of terms, isotropic and anisotropic. The anisotropic terms lead to long-range forces and their effects have not yet been pursued very far (the constraint tensor is a marginal operator). By contrast, the effects of the isotropic terms have been studied through the renormalization group (Wegner, 1974) with the following results, obtained for zero pressure. If without the coupling the specific heat diverges ($\alpha > 0$), then with the coupling the transition becomes first order. Otherwise ($\alpha < 0$), for weak coupling the transition is second order with unrenormalized exponents; for strong coupling it is first order; and for a certain intermediate value of the coupling a kind of tricritical point appears, with Fisher-type-renormalized exponents (but renormalized in the opposite direction since α is negative).

Balance sheet

The theory of magnetoelastic couplings is in a state which we must hope is provisional. One knows enough to list the possible effects, but not enough to predict which effects will actually dominate in a given system. There is a wealth of experimental information, discussed in some of the theoretical papers we have quoted.

11.2 Coupling to a gauge field

In a superconductor the order parameter is coupled to the electromagnetic field; indeed superconducting order arises from the formation of charged pairs of electrons (Cooper pairs, each pair carrying double the electronic charge). The electromagnetic field has fluctuations of its own, whose influence on the super-conducting fluctuations is not negligible in principle, as we shall see.

The coupling between the superconducting order parameter and the electromagnetic field is determined by the condition that it be gauge-invariant. The Hamiltonian of an electron in an electromagnetic field is obtained by the replacement $\mathbf{p} \to \mathbf{p} - e\mathbf{A}$, where $\mathbf{p}$ is the electron momentum and $\mathbf{A}$ the electromagnetic vector potential. Again because of gauge invariance, the gradient term in the Ginzburg–Landau free energy is modified by the replacement

$$\nabla\psi \to (\nabla - iq\mathbf{A})\psi$$

where ψ is the (complex) order parameter, and $q = 2e/\hbar c$.

In this way one obtains for the local free-energy density the expression

$$F\{\psi, \mathbf{A}\} = F_0 + \tfrac{1}{2}r_0|\psi|^2 + u_0|\psi|^4 + \tfrac{1}{2}|(\nabla - iq\mathbf{A})\psi|^2 + K(\nabla \times \mathbf{A})^2$$

where the last term is the energy of the electromagnetic field.

Landau method. Elimination of electromagnetic fluctuations

We adopt an intermediate kind of Landau approach, neglecting terms with gradients of Ψ but not those with gradients of $\mathbf{A}$; (Halperin, Lubensky, 1974). Keeping ψ constant one can calculate the fluctuations of $\mathbf{A}$ since the density $F\{\psi, \mathbf{A}\}$ is quadratic in $\mathbf{A}$:

$$\langle A^2(k) \rangle \sim \frac{1}{k^2 + |\psi|^2}$$

Substituting this value into F one obtains an effective free energy as a function of $|\psi|$:

$$F(\psi) = F_0 + \tfrac{1}{2}\tilde{r}_0|\psi|^2 + u_0|\psi|^4 - Cq^2|\psi|^d$$

The parameter r_0 is slightly changed, but the important new feature is the last term, written above in a form that is correct for $2 < d < 4$. In the framework of the Landau approximation the effect of this term is to make the transition first order.

Renormalization-group approach

Such an approach has been made through an expansion in powers of $\varepsilon = 4 - d$, by generalizing the dimensionality n of the order parameter to arbitrary values. In the third step (see Section 4.1) one must then renormalize not only the variable $\psi(\mathbf{x})$ but also the variable $\mathbf{A}(\mathbf{x})$. The results can be visualized in the (q^2, u_0) plane of parameter space (Lubensky, 1974). The 'neutral' fixed point (at $q^2 = 0$, in absence of the interaction) is unstable for all values of n; in other words the coupling is relevant. The topology of the fixed points depends on the value of n. When $n > n_c(d)$, ($n_c(d)$ being very large, of the order 366), two new fixed points appear, one stable and the other unstable. When $n = n_c(d)$, these two fixed points coincide, but instead of exchanging stabilities in the usual manner, both become complex when $n < n_c(d)$, and thus disappear from parameter space; the trajectories then tend to infinity, which one interprets as a sign that the transition is of first order.

For $n = 2$, corresponding to superconductivity, it is very likely that $n < n_c(d)$ when $2 < d < 4$, in which case one recovers the predictions of the preceding subsection.

Discussion

The presence of such an interaction differentiates superconductivity from superfluidity in ^{4}He, though both have order parameters with the same dimensionality $n = 2$. In practice the effects of the interaction, and especially the discontinuity in the order parameter, are probably very small, because the range of the forces causes the critical region in superconductors to be very narrow. Such couplings make themselves felt also in smectic-A to nematic transitions, which are formally analogous to the superconducting transition (de Gennes, 1972); the vector potential is replaced by the director **n**, and electromagnetic fluctuations by fluctuations of molecular orientation. But one expects more pronounced effects than in superconductors because the critical region is wider (Halperin, Lubensky, 1974; Halperin, Lubensky, Ma 1974).

It is possible to gain some further physical insight into the coupling mechanism. For superconductors this mechanism is the Meissner effect, i.e. the expulsion of the magnetic field. In more general terms this is a manifestation of the Higgs mechanism (Abers, Lee, 1973); the transverse vectorial electromagnetic mode (i.e. the photon, a gauge field with zero mass) couples to a scalar mode, namely the phase of the superconducting order parameter (Goldstone mode with zero mass), giving a composite vectorial mode which has finite mass as consequence of the coupling. As a result the fluctuations are inhibited, and here this is reflected by the transition becoming first order.

11.3 Impurity effects

Impurity effects plainly span an enormous range of problems, which can be viewed either as couplings to other degrees of freedom or as consequences of inhomogeneities. This is one way of interpreting the distinction drawn in Section 7.4 between the problems of mobile as opposed to frozen impurities.

We shall start with the simpler problem of mobile impurities, by discussing an exactly soluble model due to Syozi; this is an Ising model with non-magnetic impurities. Then we shall present the results obtained by the renormalization-group method for the problem of frozen impurities.

The Syozi model

This is one of the so-called 'decorated' models (Syozi, 1972). One starts with magnetic atoms on a lattice, and an Ising-type ferromagnetic interaction. Then one introduces non-magnetic impurities on the bonds between lattice sites, whose

154

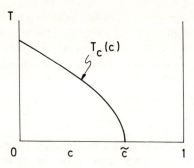

Figure 11.1. Temperature vs. concentration diagram for impurities, showing the domain where the ferromagnetic phase exists in Syozi's model

effect is to decouple the two adjacent sites. At constant chemical potential of the impurities, the partition function is then analogous to that of the unperturbed Ising model. Imposing a prescribed concentration of impurities amounts to imposing a global constraint in the sense of Section 7.3, and entails a Fisher-type renormalization of the exponents.

Figure 11.1 sketches the variation of the critical temperature as a function of the impurity concentration; the former drops as the latter rises. There exists a value $\tilde{c}$ of the concentration at which the critical temperature is reduced to zero. On the other hand, when non-magnetic impurities are distributed on bonds at random as in the percolation problem (Appendix 2.2), we know that there exists a concentration c_0 at which infinite clusters of unobstructed bonds cease to occur. It is interesting to compare the values of $\tilde{c}$ and of c_0. One finds in fact the result

$$\tilde{c} > c_0$$

which can be understood as a consequence of the correlations between the positions of the impurities when these are mobile. Such correlations allow ferromagnetism to survive over a wider domain, by adjusting positions to the ferromagnetic order. This result has conceptual importance, for it illustrates clearly the differences between problems with mobile and frozen impurities, respectively.

At low concentrations (c small) one has an exponent $\phi = \alpha$,

$$(\Delta T)^* \sim c^{1/\alpha}$$

which determines the width of the renormalized critical region, and an exponent $\psi = 1$,

$$T_c(0) - T_c(c) \sim c$$

which determines the shift in critical temperature. Note that for the Ising model one has $0 \leqslant \alpha < 1$.

Frozen impurities

The general problem of frozen-in disorder, and its effects on critical phenomena, can be studied in the Ginzburg–Landau approach by the method of the renormalization group (Lubensky, 1975). For an inhomogeneous system the partition function Z is

$$Z = \int \int \prod_x \{d^m \mathbf{M}(\mathbf{x})\} \exp(-\mathcal{H}),$$

where

$$\mathcal{H} = \int d^d\mathbf{x} \{A(\mathbf{x})|\mathbf{M}(\mathbf{x})|^2 + V|\mathbf{M}(\mathbf{x})|^4\}$$

$M(x)$ is an m-component vector. 'Disorder' is introduced by replacing $\mathbf{A}(\mathbf{x})$ by $\bar{A}(\mathbf{x})$ $+ \psi(\mathbf{x})$, where $\psi(\mathbf{x})$ is a random variable representing, say, disorder in the short-range exchange interaction. When the disorder is frozen, one starts by calculating the free energy F_i for a disordered configuration $\{i\}$, and then one averages over all configurations:

$$F = \langle F \rangle_{\{i\}}$$

(By contrast, for mobile impurities the average is taken directly in the partition function Z.) In order to calculate $\langle F \rangle$ one uses the following trick:

$$-\frac{F}{kT} = -\frac{1}{kT} \langle F \rangle_{\{\psi\}} = \langle \log Z \rangle_{\{\psi\}} = \frac{\partial}{\partial n} \langle Z^n \rangle_{\{\psi\}}|_{n=0}$$

which leads one to introduce a variable $\boldsymbol{\sigma}$ having $n \times m$ components $[\sigma = (\mathbf{M}_1, \dots \mathbf{M}_n)]$. Then Z^n is written as

$$Z^n = \int \int \prod_x d^{nm} \boldsymbol{\sigma}(\mathbf{x}) \exp[-\mathcal{H}(\boldsymbol{\sigma}, \psi)]$$

where

$$\mathcal{H}(\boldsymbol{\sigma}, \psi) = \int d^d\mathbf{x} \left\{ (\bar{A}(\mathbf{x}) + \psi(\mathbf{x}))|\boldsymbol{\sigma}(\mathbf{x})|^2 + V \sum_{\alpha=1}^{n} |\mathbf{M}_\alpha(x)|^4 \right\}$$

If one now assumes that the distributions of $\psi(\mathbf{x})$ and of $\psi(\mathbf{x}')$ are mutually independent, then one is led to

$$\langle Z^n \rangle_{\{\psi\}} = \int \int \prod_x d^{nm} \boldsymbol{\sigma}(\mathbf{x}) \exp[-\mathcal{H}_{\text{eff}}(\boldsymbol{\sigma})] = Z_{\text{eff}}$$

where the effective 'Hamiltonian', invariant under translations, is

$$\mathcal{H}_{\text{eff}}(\boldsymbol{\sigma}) = \int d^d\mathbf{x} \left[\bar{A}(\mathbf{x})|\boldsymbol{\sigma}(\mathbf{x})|^2 + V \sum_{\alpha=1}^{n} |\mathbf{M}_\alpha(\mathbf{x})|^4 + u|\boldsymbol{\sigma}(\mathbf{x})|^4 \right]$$

with

$$u = -\tfrac{1}{2}\langle \psi^2 \rangle, \qquad (\langle \psi \rangle = 0)$$

Hence the free energy of the disordered system is given by

$$F = \left.\frac{F_{\text{eff}}}{n}\right|_{n \to 0} = \left. -kT\frac{\partial}{\partial n}Z_{\text{eff}}\right|_{n=0}$$

The effective 'Hamiltonian' is of the same kind as the one introduced in Chapter 9 to study cubic anisotropy. The renormalization-group method (expansion in the vicinity of $d = 4$) yields the asymptotic critical behaviour, which depends on the number m of components of the order parameter. With m_α defined as the value of m for which $\alpha = 0$, there is a new kind of critical behaviour, having its own characteristic critical exponents, when $1 < m < m_\alpha$, while for $m > m_\alpha$ the disorder fails to affect the critical behaviour.

11.4 Conclusion

The problems so briefly surveyed in this chapter call for much theoretical and experimental elaboration. In concluding one should mention the possible similarities of the problems raised in Section 11.2 to the unified theories of weak and electromagnetic interactions, (Weinberg, 1974).

References

Abers, E. S., Lee, B. W. (1973), *Phys. Reports 9C*, No. 1.
Anderson, P. W. (1973), in *Collective Properties of Physical Systems*, Nobel Symposium XXIV, Academic Press.
Baker, G. A., Essam, J. W. (1970), *Phys. Rev. Letters*, **24**, 447; (1971), *J. Chem. Phys.*, **55**, 861.
de Gennes, P. G. (1972), *Solid State Comm.*, **10**, 753.
Gunther, L., Bergman, D. J., Imry, Y. (1971), *Phys. Rev. Letters*, **27**, 558.
Halperin, B. I., Lubensky, T. C., Ma, S. K. (1974), *Phys. Rev. Letters*, **32**, 292.
Halperin, B. I., Lubensky, T. C. (1974), *Solid State Comm.*, **14**, 997.
Lubensky, T. C., *Private Communication*, 1974.
Lubensky, T. C. (1975), *Phys. Rev.*, B **11**, 3219.
Pincus, P. (1971), *Solid State Comm.*, **9**, 1971.
Syozi, I. (1972), in *Phase Transitions and Critical Phenomena*, vol 1, Academic Press.
Wagner, H., Swift, J. (1970), *Z. Physik*, **239**, 182.
Wegner, F. J. (1974), *J. Phys. C*, 7, 2109.
Weinberg, S. (1974), *Sci. Am.*, **231**, No. 1; *Rev. Mod. Phys.*, **46**, 255.

CHAPTER 12
Tricritical points

12.1 Introduction

We have already, in several different contexts, encountered the concept of a tricritical point. In the present chapter we discuss it more systematically and generalize to polycritical points, of which ordinary and tricritical points are the simplest special cases. It should be remembered that the systematic study of tricritical points started only a few years ago, and that both theoretically and experimentally it is still in full swing.

As its name is meant to suggest (Griffiths, 1970) a tricritical point is a point where three critical lines meet. To visualize a typical phase diagram we consider in addition to the temperature T and the magnetic field H (coupled to the order parameter M), another field h (coupled to a variable m). In the plane $H = 0$, there is a line L_1 of critical points, terminating at the tricritical point P_t (see Figure 12.1); the region bounded by the line L_1 is a surface of first-order transitions (involving a discontinuity in the order parameter). Out of the plane $H = 0$, there exist two further first-order surfaces disposed symmetrically like two wings, and intersecting the plane $H = 0$ along a line L of first-order transitions; (L, a continuation of the

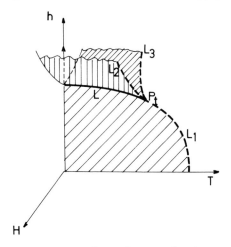

Figure 12.1. Phase diagram showing a tricritical point P_t. The cross-hatched surfaces are surfaces of first-order transitions; the line L is a line of first-order transitions; the broken lines L_1, L_2, L_3 are lines of second-order transitions

157

158

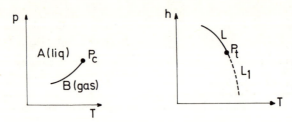

Figure 12.2. Phase diagrams illustrating the analogy
between a critical point P_c and a tricritical point P_t

line L_1, is in some sense a line of triple points). Thus, in the parameter space (H, h, T) there exist three first-order surfaces; they are bounded by three lines L_1, L_2, L_3 of critical points, which meet at the tricritical point P_t, whence issues a line L of first-order transitions formed by the intersection of the three first-order surfaces.

Compare this situation to an ordinary critical point, where two phases A and B of a system become identical. For instance in a liquid-to-gas transition, whose coexistence curve is shown in Figure 12.2, the densities ρ_L and ρ_G of the liquid and gas become equal at the critical point P_c. Similarly we can consider a tricritical point as a point where three phases A, B and C become identical. Thus, in an enlarged parameter space (admitting $h \neq 0$), the coexistence line and the critical point are replaced, respectively, by three first-order surfaces (coexistence surfaces for A and B, B and C, or C and A), and by three critical lines (loci where A = B, B = C, or C = A); finally the tricritical point is the point where A = B = C.

Admittedly there persists a certain ambiguity in the definition of a tricritical point. Depending on which physical parameter space we choose, a tricritical point can be thought of either as the point where a line of critical points changes into a line of first-order transitions, or as the meeting point of three critical lines, i.e. as a 'critical' point of a system with three components (or with three phases). As we shall have occasion to discuss later in this chapter, the two definitions are not always equivalent.

A tricritical point differs from an ordinary critical point in the number of relevant fields. It has an additional field h. At points on the critical line L_1 the fluctuations in the order parameter M diverge, inducing fluctuations in the variable m conjugate to h. Consequently the effect of the field h is to shift the critical point in question but without changing its character, except in the case of the tricritical point, where there is competition between the fluctuations of m and of M. Beyond the tricritical point any small variation in h inhibits the fluctuations, and results in a first-order transition with a discontinuity in the order parameter. Hence such competition leads to crossover behaviour which we shall study in Sections 12.2 and 12.3, first in the context of homogeneity properties, and then by the renormalization-group approach. The change from critical to tricritical point resulting from the introduction of an additional field h can be generalized; thus we shall show in Sections 12.3 and 12.4 how polycritical points can be defined. Section 12.5 will discuss the experimental situation.

12.2 Scaling

For simplicity we consider the homogeneity properties of only the thermodynamic variables. In the neighbourhood of the tricritical point $P_t(h_t, T_t)$ we introduce in addition to the field H two scaling fields μ_1 and μ_2. Then the tricritical point is fixed by the three conditions $H = 0, \mu_i = 0, \mu_2 = 0$. As for an ordinary critical point, it is natural to choose one of the two axes μ_1, μ_2 tangent to the transition line (see Figure 12.3), since this direction will play a special role. The homogeneity hypothesis for the free-energy density G is then written (Riedel, 1972; Griffiths, 1973).

$$G(H, \mu_1, \mu_2) \sim s^{-d} G(s^{y_h} H, s^{y_I} \mu_1, s^{y_{II}} \mu_2) \tag{12.1}$$

where y_h is the dimension of H and y_I, y_{II} the two next-highest and positive dimensions, which we associate with μ_1 and μ_2: ($y_{II} < y_I < y_h$, and $y_{III} < 0$, by assumption). Then μ_1, μ_2 and H are the only relevant fields; (at an ordinary critical point $y_{III} < 0$, and $y_I = \dfrac{1}{v}, \dfrac{y_h}{y_I} = \Delta$). At the tricritical point we define

'tricritical' exponents $\alpha_t, \beta_t, \gamma_t, \ldots, \left(y_I = \dfrac{1}{v}, \dfrac{y_h}{y_I} = \Delta_t \right)$, plus the crossover exponent $\phi_t = \dfrac{y_{II}}{y_I}, (\phi_t < 1)$. Accordingly, a tricritical point is characterized by three exponents, namely by y_h, y_I, y_{II}, or by v_t, Δ_t, ϕ_t.

The critical line

We have assumed that the lines L and L_1 (Figure 12.1) in the plane $H = 0$ have a common tangent, the μ_2-axis being chosen in its direction (Figure 12.3). By virtue of the homogeneity rule (12.1), the equation of the transition line near the tricritical point is

$$\mu_{1C} \sim \mu_{2C}^{1/\phi_t} \tag{12.2}$$

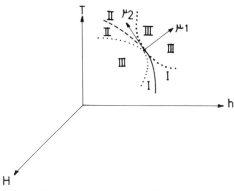

Figure 12.3.   Schematic phase diagram near the tricritical point; the crossover lines, shown dotted, separate the three different regions I, II, III

(which amounts to saying that the shift exponent defined in Chapter 8 is equal to the crossover exponent ϕ_t).

Approach to the tricritical point

The critical behaviour depends on whether the tricritical point is approached tangentially along the transition line, or from some other direction.

Non-tangential approach. If the tricritical point is approached along the μ_1 axis ($\mu_2 = 0$), then (12.1) takes the form:

$$G \sim \mu_1^{2 - \alpha_t} G\left(\frac{H}{\mu_1^{\Delta_t}}, 1, 0\right) \tag{12.3}$$

where we have assumed $2 - \alpha_t = \nu_t d$; this is just the usual critical behaviour with the 'tricritical' exponents $\alpha_t, \beta_t, \ldots$, interrelated by the scaling laws given in Section 3.1. The scaling field μ_1 is a function both of $\Delta T = T - T_t$ and of $\Delta h = h - h_t$, which therefore play analogous roles and generate similar singularities, $\left(\frac{\partial m}{\partial h} \sim \frac{\partial S}{\partial T} \sim \Delta T^{-\alpha_t} \sim \Delta h^{-\alpha_t}\right)$.

Tangential approach. If the tricritical point is approached along the μ_2 axis ($\mu_1 = 0$), then (12.1) takes the form

$$G \sim \mu_2^{(2 - \alpha_t)/\phi_t} G\left(\frac{H}{\mu_2^{(\Delta_t/\phi_t)}}, 0, 1\right) \tag{12.4}$$

Hence the critical exponents are renormalized by the crossover exponent ϕ_t. This is the generalization to three fields of the renormalization which, for an ordinary critical point, leads from the behaviour $M \sim (\Delta T)^\beta$ at constant field ($H = 0$) to the behaviour $M \sim H^{1/\delta}$ (with $1/\delta = \beta/\Delta$) when H varies at fixed $T = T_c$.

Crossover behaviour

The presence of two relevant fields μ_1 and μ_2 competing near the tricritical point induces crossover behaviour, like that described in Chapter 8 in the context of quadratic anisotropy (μ_1, μ_2, ϕ_t now replace t, Δ, ϕ). Because there now exist, simultaneously, a line of critical points and a line of first-order transitions, we must distinguish as in Figure 12.3 between three regions I, II, III separated by crossover lines having equations $\mu_1^* \sim \mu_2^{*(1/\phi_t)}$; each region corresponds to a different regime (first order, critical and tricritical transitions). We shall consider in turn the crossover from III to II and from III to I. When μ_2 is fixed and μ_1 large enough, tricritical behaviour is observed; but for $\mu_1 \ll |\mu_2|^{1/\phi_t}$ one meets, either a new kind of critical behaviour if $\mu_2 > 0$, or first-order transitions if $\mu_2 < 0$.

Tricritical to critical crossover. Consider for instance the susceptibility $\chi = -\dfrac{\partial^2 G}{\partial H^2}$; the same general argument applies to any variable having a critical singularity, thus for instance to $\partial m/\partial h$. The homogeneity hypothesis (12.1) becomes

$$\chi \sim \mu_1^{-\gamma_t} f\left(\frac{\mu_2}{\mu_1^{\phi_t}}\right) \tag{12.5}$$

As long as $\mu_1 \gg \mu_2^{1/\phi_t}$, in the tricritical region III, one has tricritical behaviour with exponents $\alpha_t, \beta_t, \gamma_t$. When $\mu_1 \ll \mu_2^{1/\phi_t}$, in the critical region II, there is a new kind of critical behaviour with exponents α, β, γ, and the susceptibility χ varies like

$$\chi \sim (\mu_1 - \mu_{1c}(\mu_2))^{-\gamma} \mu_2^{(\gamma - \gamma_t)/\phi_t} \tag{12.6}$$

Tricritical to first-order crossover. Fix μ_2 at a small negative value, and vary μ_1 so that it approaches the line of first-order transitions. Consider for instance the quantity $m - m_t = \Delta m$, which according to (12.1) behaves like

$$\Delta m \sim \mu_1^{1-\alpha_t} g\left(\frac{\mu_2}{\mu_1^{\phi_t}}\right) \tag{12.7}$$

In the tricritical region III one has $\Delta m \sim \mu_1^{1-\alpha_t}$, while in region I Δm remains finite and as a function of μ_2 varies like

$$\Delta m \sim \mu_2^{(1-\alpha_t)/\phi_t} \tag{12.8}$$

which is similar to the behaviour of the transverse susceptibility $\chi_\perp$ under quadratic anisotropy (Chapter 8).

If one thinks of the tricritical point as one endpoint of the line of first-order transitions in the (μ_1, μ_2) plane, then by analogy to ordinary critical points one can define so-called 'subsidiary' exponents (Griffiths, 1973): $2 - \alpha_u = \dfrac{2 - \alpha_t}{\phi_t}$, $\gamma_u = \dfrac{\alpha_t}{\phi_t}$, $\beta_u = \dfrac{1 - \alpha_t}{\phi_t}$, $\delta_u = \dfrac{1}{1 - \alpha_t}$, $\Delta_u = \dfrac{1}{\phi_t}$; (see Equation 12.7). These subsidiary exponents, which are interrelated by the scaling laws, are not new, being expressible as functions of the tricritical exponents α_t and ϕ_t.

Conclusion

The neighbourhood of a tricritical point, like that of an ordinary critical point, accommodates scaling properties which can be exhibited by choosing two scaling fields μ_1 and μ_2, different from the physical fields T and h, and one of which corresponds to the direction tangential to the transition line. With the tricritical point we associate critical exponents three of which are independent (e.g. α_t, γ_t and ϕ_t), the others being deducible by appeal to the scaling laws. The additional crossover exponent, ϕ_t, stemming from the presence of an additional relevant field, enters in several ways. First, it is involved directly in the equation for the critical line near the tricritical point. Next, it affects the renormalization of the tricritical exponents when the tricritical point is approached tangentially. Finally it specifies the crossover lines separating the regions III, II, I, which correspond, respectively, to the tricritical regime (with exponents $\alpha_t, \beta_t, \ldots$), to the critical regime (with exponents α, β, γ, assumed to be constant on the three critical lines), and to first-order transitions (with discontinuities in the various functions).

12.3 Renormalization-group approach

As in the case of ordinary critical points, this method allows one to verify the homogeneity hypotheses adopted in the last section, and at least under some conditions to calculate the exponents and the scaling functions.

Landau theory. Tricritical and polycritical points associated with Gaussian fixed points

Start by recalling (Section 2.5) how Landau theory can describe a simple situation involving a tricritical point. Consider a local free-energy density of the form

$$F_L - F_0 = AM^2 + B_4M^4 + B_6M^6 + \text{gradient terms} \tag{12.9}$$

In Landau theory, an ordinary critical point corresponds to a zero of A, while B_4 is positive. When $B_4 < 0$, the transition is first order. If by varying the physical parameters one can cause B_4 and A to vanish simultaneously while $B_6 > 0$, then one obtains a tricritical point, understood as a point at the junction of a line of critical points ($B_4 > 0$) with a line of first-order transitions ($B_4 < 0$). In Landau theory the tricritical exponents are easy to find. While v, η, γ have the same values as at an ordinary critical point (since they follow from the Gaussian correlation function), the exponent β becomes $\frac{1}{4}$ (instead of $\frac{1}{2}$), and $2 - \alpha$ becomes $\frac{3}{2}$ (instead of 2). Then the characteristic dimensionality for which Josephson's law holds is $d_c = 3$, and one expects that for this kind of tricritical point, and in the usual three-dimensional cases, Landau theory will be correct up to logarithmic corrections.

When the transformations of the renormalization group (Chapter 4) are applied to the effective 'Hamiltonian' $\mathcal{H}$ constructed from (12.9), the recursion relations for A, B_4 and B_6 show that, for $d = d_c = 3$, the most stable fixed point is the Gaussian. In its neighbourhood, the fields conjugate to the scaling operators $O_{m,\ell}$ of degree $(2m + \ell)$ in M have the dimension (see Sections 5.2 and 7.2)

$$y_{m,\ell} = 3 - \frac{(2m + \ell)}{2} \tag{12.10}$$

Two isotropic operators (with $\ell = 0$) are relevant, $O_{1,0}$ and $O_{2,0}$, while the marginal operator $O_{3,0}$ entails logarithmic corrections to the power laws (see Section 5.4). Thus the dimensionality 3 marks the change in stability of the Gaussian fixed point with respect to the field conjugate to $O_{3,0}$. The renormalization-group approach as described in Chapter 4 then allows one to derive the scaling property (12.1), the values appropriate to $d = d_c = 3$ being $y_h = y_{0,1} = \frac{5}{2}, y_I = y_{1,0} = 2$ and $y_{II} = y_{2,0} = 1$, which entails $v_t = \frac{1}{2}, \Delta_t = \frac{5}{4}$ and $\phi_t = \frac{1}{2}$.

The arguments above generalize for a local free-energy density of the form

$$F_L - F_0 = AM^2 + B_4M^4 + \cdots + B_pM^p + \text{gradient terms} \tag{12.11}$$

In Landau theory, a zero of A at a point where $B_4 = B_6 = \cdots = 0$, while $B_p > 0$, corresponds to a polycritical point, often called a critical point of order $\frac{p}{2}$; (in this language an ordinary critical point is of order 2 and a tricritical point is of order 3).

Such a point can be defined as the junction of a line of critical points of order $\frac{p}{2} - 1$, $(B_{p-2} > 0)$, and a line of first-order transitions, $(B_{p-2} < 0)$. There exists a characteristic dimensionality $d_c = \frac{2p}{p-2}$ such that for $d < d_c$ the exponents are given by Landau theory. When $d = d_c$, the renormalization-group approach associates the polycritical point with a Gaussian fixed point to which $\frac{p}{2} - 1$ isotropic operators are relevant, and the operator M^p is marginal. Then the expression (12.10) generalizes to

$$y_{m,l} = d_c - (2m + l)\left(\frac{d_c - 2}{2}\right) \tag{12.12}$$

Thus the critical behaviour near a polycritical point of order $\frac{p}{2}$ is governed by $\frac{p}{2}$ independent critical exponents, namely the usual two critical exponents plus $\frac{p}{2} - 2$ crossover exponents. They are easily expressed in terms of the dimensions $y_h, y_{\rm I}, y_{\rm II}$, etc.

The crossover behaviour described in Section 12.2 can be visualized more clearly by following the renormalization-group method and considering the renormalization trajectories as shown in Figure 12.4. We assume $d = d_c\left(\frac{p}{2}\right)$. At the polycritical point $\left(\text{of order } \frac{p}{2}\right)$ the renormalization trajectory issues from the physical point A, and travels to the Gaussian fixed point $P^*_{G,p/2}$. Point B, on the

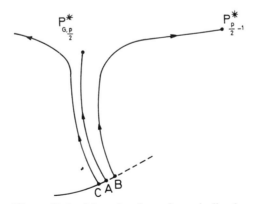

Figure 12.4. Map showing schematically the fixed points and the pattern of trajectories for a critical point of order $p/2$, at the characteristic dimensionality $d = d_c(p/2)$

line of critical points of order $\frac{p}{2} - 1$, is very close to A, and the trajectory issuing from B travels at first towards the fixed point $P^*_{G,p/2}$, but then veers away to the non-Gaussian fixed point $P^*_{p/2-1}$;(since $d = d_c\left(\frac{p}{2}\right) < d_c\left(\frac{p}{2} - 1\right)$). The trajectory issuing from the point C, on the line of first-order transitions and also close to the polycritical point, likewise travels at first towards the Gaussian fixed point before veering off to infinity.

Tricritical points associated with a non-Gaussian fixed point

Start by considering a tricritical point obtained from the Expression (12.9) for the free-energy density. For dimensionalities d below the characteristic dimensionality $d_c = 3$, the Gaussian is no longer the most stable fixed point; there exists another one nearby which governs the tricritical behaviour. This new non-Gaussian fixed point coalesces with the Gaussian when $d = d_c = 3$. As in the case of an ordinary critical point, it is possible to determine the new fixed point and its critical exponents by expansions in powers of $\varepsilon = d_c - d = 3 - d$; (Stephen, McCauley, 1973). The exponents γ_t and η_t, (for arbitrary n, and to order ε^2), and ϕ_t (for $n = 1$, and to order ε) are

$$\gamma_t = 1 + \frac{5}{8}\frac{(n + 2)(n + 4)}{(3n + 22)^2}\varepsilon^2$$

$$\eta_t = \frac{1}{12}\frac{(n + 2)(n + 4)}{(3n + 22)^2}\varepsilon^2 \qquad (12.13)$$

$$\phi_t = \frac{1}{2} + \frac{1}{10}\varepsilon$$

These considerations are easily generalized to a polycritical point $\left(\text{of order } \frac{p}{2}\right)$, and the critical exponents can in principle be found from expansions in powers of $\varepsilon = d_c\left(\frac{p}{2}\right) - d, (d < d_c)$.

Polycritical points occur also in different kinds of situations; one example is the tricritical point which can appear in the presence of cubic anisotropy, as already mentioned in Chapter 9. Then, for $n > n_c(d)$, a tricritical point can be associated with the isotropic fixed point, with the critical exponents corresponding; the crossover exponent ϕ_t is associated with the dimension y_v of the cubic anisotropy field v. If v is small and positive, the renormalization trajectory first approaches the isotropic fixed point, and then veers away towards the cubic fixed point which is the more stable of the two; this results in second-order transitions. If v is negative, then the trajectory travels out to infinity, resulting in a first-order transition. Thus, for $v = 0$ there exists a tricritical point at the junction of a line of critical points (for $v > 0$) with a line of first-order transitions.

Conclusion

In the language of the renormalization group, we say that a fixed point corresponds to a tricritical point under the following conditions: on the critical surface, the result of any slight variation of a parameter is, according to the sign of the variation, either that the critical trajectory veers away to another more stable fixed point, or that the trajectory travels out to infinity as an indication that the transition has become first order. Thus a tricritical point is defined as the junction between a critical line and a line of first-order transitions. By studying the neighbourhood of a tricritical fixed point, we can in principle determine simultaneously the tricritical exponents, the scaling fields μ_1 and μ_2 discussed in Section 12.2, and the scaling functions describing the crossover.

12.4 Polycritical points

We revert for a moment to tricritical points. In ternary or quaternary fluid mixtures there exist points of parameter space where three phases that are present become identical simultaneously. These are tricritical points representing generalizations of the Landau tricritical point, the latter being just the special case (Griffiths, 1974) so far dealt with in this chapter. Hence it is convenient to define a tricritical point as a triple critical point where three phases that are present become identical. A natural generalization considers four, five or more phases becoming identical simultaneously; for instance in a system with five components one expects to observe a critical point where four phases become identical. The polycritical points introduced in Section 12.3 by aid of the generalized Ginzburg–Landau formulation (12.11) correspond to just this situation; (they terminate a line of first-order transitions which can be considered as a multiple line).

However, there exist, too, critical points which are located at the junction of several transition lines and which are not readily assigned to any of the foregoing categories. The 'spin-flop' transitions in certain magnetic materials are one example. In an n-component antiferromagnetic with uniaxial anisotropy, there are certain values of n such that the application of a uniform magnetic field h along the anisotropy axis leads to the phase diagrams shown in Figure 12.5; (Fisher and Nelson, 1974, have studied this model by renormalization-group methods). Two lines of critical points (with different exponents) meet at the point b. This point, called 'bicritical', (and possessing its own set of critical exponents), is the junction of two lines of critical points, and of one first-order transition line which separates an antiferromagnetic from a so-called 'spin-flop' phase; in the latter the spins have rotated so as to acquire a non-zero component perpendicular to the aniso-tropy axis. For certain values of n, the first-order line is replaced by two lines of critical points with a new intermediate phase included between them; in that case there are four critical lines meeting at b, which is then called a 'tetracritical' point. This is a good example of the great diversity of possible phase diagrams, and of the need for a systematic classification.

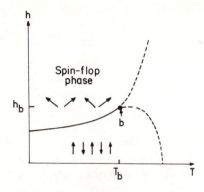

Figure 12.5. Schematic phase diagram for an anisotropic anti-ferromagnet, displaying a 'bicritical' point in a uniform parallel field h. The solid line is a line of first-order transitions; the broken lines are lines of second-order transitions

12.5 Experimental results

Points where variation of a parameter changes a transition from first to second order are known in ^{3}He + ^{4}He mixtures, metamagnetic solids and certain ammonium halides. But only the helium mixtures have been investigated in enough detail to illustrate the theoretical predictions made in the preceding sections. In the other two systems, whose study is less far advanced, one encounters difficulties when trying to identify and locate a tricritical point correctly.

^{3}He–^{4}He mixtures

On mixing two liquids like ^{3}He and ^{4}He one obtains phase-separation curves in the (T, X) phase as shown in Figure 12.6; here X is the concentration or molar fraction of ^{3}He. But at low temperatures ^{4}He becomes superfluid at a critical point, often called the λ-point because of the graphical appearance of the specific heat singularity. In the presence of ^{3}He, the λ-point moves downwards to meet the rising phase-separation curve, resulting in the phase diagram shown in Figure 12.7; (there is compensation between energy gain in the superfluid state and energy loss from the heat of mixing). Thus the critical point at the peak of the phase-separation curve is a tricritical point. Reverting to the notation of Section 12.1, M is the superfluid order-parameter; the field H conjugate to M is not observable experimentally; m is the concentration X, and the field b conjugate to m is the difference $\mu_3 - \mu_4$ between the ^{3}He and the ^{4}He chemical potentials, which is accessible, indirectly, to experiment. Then the phase diagram in the (b, T) plane is analogous to Figure 12.2.

This system can be represented by the Ginzburg–Landau model discussed in Section 12.3, which for $d = 3$ has a tricritical point with classical exponents. The

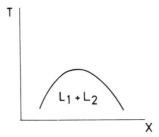

Figure 12.6. Phase-separation curve for a mixture of
two liquids

exponents have been measured experimentally and serve to confirm this predic-
tion. For instance, the 'susceptibility' $\dfrac{\partial X}{\partial(\mu_3 - \mu_4)} = \dfrac{\partial m}{\partial h}$ has been found at
constant concentration, [which in the (h, T) plane corresponds to a tangential
approach to the tricritical point]. The asymptotic behaviour of this function
agrees perfectly with theory (Goellner, Meyer, 1971):

$$\frac{\partial X}{\partial(\mu_3 - \mu_4)} \sim (T - T_t)^{-1,0}$$

where T_t is the tricritical temperature. (Landau theory predicts $\partial m/\partial h \sim$
$(T - T_t)^{-(\alpha_t/\phi_t)}$, with $\alpha_t = \frac{1}{2}$ and $\phi_t = \frac{1}{2}$.)

Metamagnetic systems

Metamagnetic systems, $FeCl_2$ for instance, generally consist of planes; the
magnetic ions are coupled through an interaction that is anisotropic (both in spin
space and in real space), and is ferromagnetic within each plane but
antiferromagnetic between different planes. One should note the difference

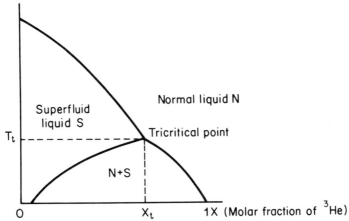

Figure 12.7. Phase diagram in the (x, T) plane for a ^3He–^4He mixture

between these systems and antiferromagnetics undergoing spin-flop transitions as discussed in Section 12.4. In the absence of any uniform magnetic field h, there is a second-order paramagnetic to antiferromagnetic transition, whose order parameter M is the alternating magnetization of the individual planes, the conjugate of M being a field H alternating in the same way. In the presence of a uniform field h one obtains the phase diagram shown in Figure 1.3 (where H should be replaced by h to accord with our present notation). As yet the experimental study of the resulting tricritical point has not advanced very far. In fact, as a consequence of the various anisotropies, a uniform applied field h can give rise to an alternating field H inside the system, thus preventing access to the tricritical point which by symmetry is located on the plane $H = 0$. However, even in this situation one can demonstrate the existence of the critical lines located on the wings (i.e. the lines L_2 and L_3 in Figure 12.1).

Hence, in the case of these metamagnetic systems where classical tricritical behaviour is expected, experimental confirmation has not yet been forthcoming.

Ammonium halides (NH_4Cl)

In these lattices two different positions are available for the ammonium tetrahedra, whence the possibility of an order–disorder transition. At ordinary pressures ammonium chloride undergoes a first-order transition; as the pressure rises this becomes second-order at a point which was originally believed to be a tricritical point. But the interpretation remains controversial; some experiments suggest that there are fluctuations associated with an antiferromagnetic order m (conjugate to a field h). It would then be possible that in an extended phase diagram (h, H, T) the point originally considered as tricritical appears as a critical point of higher order. Accurate measurements of the critical exponents should resolve this question: another good illustration of the importance of measuring exponents.

12.6 Conclusion

In conclusion we stress the desirability of constructing, eventually, a simple, systematic and complete classification of polycritical points and of intersections between transition lines. Such a classification would enable one to establish analogies between systems that are different at first sight but whose structures are similar. In this chapter we have taken only a brief look at the experimental and theoretical attempts, but in any case the work is still very much in progress. Especially interesting in such studies is the light they cast on the general problems of competition between different kinds of order, and of couplings between different degrees of freedom (see Chapter 11).

References

Fisher, M. E., Nelson, D. R. (1974), *Phys. Rev. Letters,* **32,** 1350.
Goellner, G., Meyer, H. (1971), *Phys. Rev. Letters,* **26,** 1543.

Griffiths, R. B. (1970), *Phys. Rev. Letters*, **24,** 715.
Griffiths, R. B. (1973), *Phys. Rev.*, **B7,** 545.
Griffiths, R. B. (1974), *J. Chem. Phys.*, **60,** 195.
Riedel, E. K. (1972), *Phys. Rev. Letters*, **28,** 675.
Riedel, E. K., Wegner, F. J. (1972), *Phys. Rev. Letters*, **29,** 349.
Stephen, M. J., McCauley, J. L. (1973), *Phys. Letters*, **44**A, 89.

CHAPTER 13
Marginality

'Or do you not know that between two opposites there exists a mean?'

Plato

13.1 Introduction

Marginal fields have been encountered repeatedly in earlier chapters. A field marginal with respect to a given fixed point is one whose anomalous dimension, defined in the neighbourhood of that fixed point, is zero. Marginality, the case intermediate between relevance and irrelevance, can introduce various peculiarities into critical regimes; indeed it is marginality which is responsible for the pathological features of several important cases. Hence we show in this chapter how the renormalization-group approach enables one to understand not only the rule but also the exceptions.

There are several different degrees of marginality. Consider the differential renormalization equation for a scaling field g_i:

$$\frac{dg_i}{dl} = y_i g_i + y_{ii} g_i^2 + y_{iii} g_i^3 + \cdots \tag{13.1}$$

In the non-linear terms of this expression we have replaced the other fields g_j by their expansions in powers of g_i; in other words (13.1) describes the projection of the renormalization trajectory onto the g_i axis. Marginality occurs when $y_i = 0$. When the higher-order coefficients are non-zero, one speaks of local marginality. Depending on how many of the consecutive lowest-order coefficients vanish, the marginality is more or less persistent. In the extreme case where all the coefficients vanish, we see that the projection of the representative point onto the g_i axis is immobilized: there exists, then, not just an isolated fixed point, but a continuous line of fixed points (see Figure 13.1). If we now consider the set of all fields rather than only one field, we see that a great variety of situations can arise: multiple marginality, surfaces of fixed points, and so on.

We have already met the phenomenon of local marginality in the case of $d = 4$, studied in Section 5.4. It corresponds to the coincidence of two fixed points and is reflected in parameter space by a point of inflexion (Figure 5.2). The same phenomenon occurs at several different characteristic dimensionalities, as we know from earlier chapters. Since in such cases one of the two fixed points is the trivial fixed point, there is a direct link between local marginality at $d = d_c$ and the possibility of expansions in powers of $d_c - d$. More generally we meet local marginality, associated with an exchange of stabilities, along boundary lines like

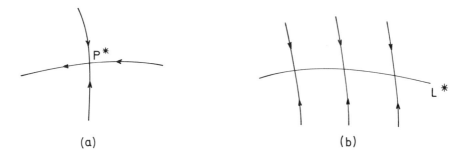

Figure 13.1. Marginal topologies. (*a*) Fixed point P* with a locally marginal field. (*b*) Line L* of fixed points associated with a persistently marginal field

those in the (n, d) table; examples are the curve L_v in the (n, d) diagram for cubic anisotropy (Figure 9.3), and the boundary lines in the (d, σ) diagram for long-range forces (Figure 10.1).

Local marginality is usually manifested by the appearance of logarithmic corrections; the mechanism underlying this effect was described in detail in Section 5.4. Such logarithmic factors constitute an evasion of the homogeneity rules.

Persistent marginality with a line of fixed points leads to more spectacular departures from universality. Critical exponents are universal only to the extent that they are associated with a unique fixed point; along a continuous line of fixed points they can vary continuously. Different values of the exponents will be observed, depending on the particular point of this line on which the trajectory converges. As a rule, variation of exponents as functions of a coupling constant stems from the existence of a line of fixed points, which in turn is due to persistent marginality of the coupling constant in question. Thus, the apparent departures from universality in the famous Baxter model (Section 9.3) can be understood in terms of the persistent marginality of cubic anisotropy.

Section 13.2 discusses various problems arising from lines of fixed points, namely short long-range order in two dimensions, and infrared anomalies in metallic spectra.

Section 13.3 is devoted to the Kondo effect, i.e. to the problem of magnetic impurities in metals. This problem, which has played an important part in the development of the renormalization group (see Section 1.2), is in some ways a jamboree of marginality. There is a formal analogy between the Kondo effect and the one-dimensional Ising model with long-range forces ($n = d = \sigma = 1$); and the former is very interesting as a model both of a certain kind of physical behaviour and of a certain method of theoretical analysis.

13.2 Lines of fixed points

Amongst the family of problems involving lines of fixed points, one should first mention the Baxter model, which is the example most often quoted, and also the

Ising model on a Bethe lattice, which surely constitutes the simplest case. But we proceed now to discuss two categories of problems that are less often considered from this angle.

Short long-range order in two dimensions

Short long-range order occurs in two-dimensional simple systems having an order parameter of dimensionality $n > 1$. The effect is linked to the existence of $(n - 1)$ angle-variables, which has a profound influence on the properties of the low-temperature phase, and which for $d = 2$ prevents the spontaneous appearance of a non-zero value of the order parameter. To understand the kind of order ('short long-range order') which can then exist, it is convenient to start by re-examining the situation for $2 < d < 4$.

For $2 < d < 4$, the transverse susceptibility in the ordered phase diverges like $1/H$ and the longitudinal susceptibility diverges like $\left(\dfrac{1}{H}\right)^{(4-\delta)/2}$; (see Section 6.3). By the usual dimensional analysis one deduces from this that the transverse and the longitudinal correlation functions behave asymptotically according to the following power laws:

$$\Gamma_T(R) \sim \frac{1}{R^{d-2}} \qquad \text{whence} \quad \eta_T = 0$$

$$\Gamma_L(R) \sim \frac{1}{R^{2(d-2)}} \qquad \text{whence} \quad \eta_L = d - 2$$

For $n = 1$, power-law behaviour obtains only at $T = T_c$, but for $n > 1$ there is critical behaviour of one kind at $T = T_c$, with a certain value η_c of the exponent η, and critical behaviour of another kind for $0 < T < T_c$, with the values η_T and η_L given above. Notice that η_c depends on n while η_T and η_L depend neither on n, nor on the temperature when $T < T_c$. These results can be summarized by saying that the critical behaviour at T_c is governed by the ordinary, non-trivial, fixed point, while the low-temperature behaviour is governed by another, trivial, fixed point; we shall call the latter the 'spin-wave fixed point', because the exponents η_T and η_L are those that are obtained in the framework of the spin-wave approximation, (in spite of the fact that otherwise this approximation is applicable only at very low temperatures).

If we now let the space-dimensionality d tend to 2, we see that $\eta_L \to \eta_T = 0$, which is consistent with the total disappearance of any spontaneous magnetization, or in other words of any distinction between the longitudinal and transverse directions. Moreover, it is believed that there appears a line of fixed points joining the ordinary and the trivial fixed points. Then the exponent η varies continuously for $0 < T < T_c$; it becomes a function $\eta(T)$ of temperature, with $\eta(0) = 0$ and $\eta(T_c) = \eta_c$. In the phase having short long-range order, the exponent δ which governs the longitudinal susceptibility, $(M \sim H^{1/\delta})$, likewise becomes a function of temperature $\left(\delta = \dfrac{4 - \eta}{\eta} \right)$.

In two dimensions and for $n > 2$, T_c is zero. But for d close to 2, T_c is small, and Brézin and Zinn-Justin (1976) have developed a method for calculating the critical exponents by expansions in powers of $\varepsilon = d - 2$; their method is inspired by low-temperature expansions, and is related to the renormalization group and to field theory.

Infrared anomalies in metallic spectra

This is a problem unrelated to phase transitions, but exhibiting scaling and universality properties. The experimental motivation stems from the interpretation of the X-ray absorption and emission spectra of metals. An X-ray transition introduces a local perturbation of the conduction electrons by creating or filling a deep hole; hence the X-ray spectrum is influenced by how these electrons return to equilibrium. The Fourier transform of the energy spectrum $I(E)$ is a function $F(t)$ of the time. Its asymptotic behaviour in a simple model has been calculated by Nozières and de Dominicis (1969); for an X-ray transition between discrete levels they find

$$F(t) \sim \left(\frac{1}{t} \right)^{(\delta_F/\pi)^2} \tag{13.2}$$

where δ_F is the phase shift (it is assumed here for simplicity that there is only one) at the Fermi level due to the local perturbation. $F(t)$ itself obeys an asymptotic power law in time, with an exponent which is universal in the sense that it depends only on the phase shift at the Fermi level. Such singular behaviour of $F(t)$, a consequence of the long time lag with which the electrons at the Fermi level return to equilibrium, entails an infrared (low energy) anomaly in the spectrum $I(E)$, likewise of power-law type.

In the language of the renormalization group, (linked here to a dilatation in time rather than in space), one can say that many of the parameters are irrelevant, (like for instance the phase shifts away from the Fermi level), but that there exists one marginal parameter, namely the phase shift at the Fermi level. Its persistent marginality entails the existence of a line of fixed points, and the continuous variation of the exponents as functions of this parameter.

The phenomenon of infrared anomalies underlies the correspondence between the Kondo effect and the one-dimensional Ising model with long-range forces (Section 13.3). Moreover there exists a remarkable isomorphism between this problem of infrared anomalies and certain simplified models of one-dimensional metals (Luttinger, 1963; Tomonaga, 1950), (see Section 13.4), and also between the former problem and a model field theory in two space-time dimensions, (Thirring, 1958).

13.3 The Kondo effect

Formulation of the problem

Consider the Hamiltonian

$$\mathscr{H} = \sum_{k\sigma} \varepsilon_k \eta_{k\sigma} - J\mathbf{S} \cdot \mathbf{s}(0) \tag{13.3}$$

where the first term is the kinetic energy of a set of free electrons, and the second term defines an exchange-interaction between the spin $\mathbf{S}$ of an impurity localized at the origin, and the spin-density $\mathbf{s}(0)$ of the conduction electrons at the impurity; the parameter J is often called the exchange integral. If J is positive, it tends to polarize the conduction electrons parallel to the impurity spin (ferromagnetic coupling); if J is negative, it tends to produce antiparallel polarization (antiferromagnetic coupling).

The Hamiltonian (13.3) thus defines a problem where electrons are scattered by a localized magnetic impurity. For a non-magnetic perturbation the problem would be very simple, because in that case the electrons scatter independently and no many-body aspects obtrude in practice; then it is only if the perturbation becomes time-dependent that one encounters the complications (infrared anomalies) discussed in the preceding section. But for a magnetic perturbation the many-body aspects of the problem cannot be evaded, because then there is an effective interaction between the electrons, due to the fact that an electron can change the state of the impurity and thereby influence the way a second electron scatters. The effective interaction thus reflects the internal structure (internal degrees of freedom) of the impurity which gives it a 'memory'. One can express the difficulty by saying that the possibility of 'rotating' the impurity spin introduces all the complications of time-dependent perturbations. In the classical limit where the magnitude $\mathbf{S}$ of the impurity spin tends to infinity ($\mathbf{S} \to \infty$, $J \to 0$, JS finite), the problem once again becomes simple, because the 'rotation' of the spin is effectively inhibited.

Accordingly, the main parameters of the problem are the exchange integral J, (or rather the dimensionless product $\tilde{J} = J\rho$ where ρ is the density of states at the Fermi level), and the value $\mathbf{S}$ of the localized spin. In the following we shall concentrate chiefly on the case $\mathbf{S} = \frac{1}{2}$; the dependence on $\mathbf{S}$ is not yet known, but can perhaps be tackled through some simple scaling argument. Other parameters are the shape of the conduction band and the position of the Fermi level; and part of the Kondo problem consists in determining to what extent the various output functions are universal, i.e. independent of the parameters.

However difficult the problem defined by the Hamiltonian (13.3), it is still only a very simplified model of impurities in a metal, lacking non-magnetic perturbations, orbital degeneracy, spin–orbit coupling, crystal fields, direct electron-lattice interactions, etc. Hence the solution of the model must be regarded as a first step; it is a reasonable supposition that one of the subsequent steps will be the solution of the Anderson model (Anderson, 1961). Afterwards one might hope to introduce the other complications in a way somewhat analogous to the systematic enlargement of parameter space in the study of phase transitions.

Parameter space

Consider the $(\tilde{J}, T)$ diagram. Because the perturbation is localized, the only temperature at which singularities can appear is absolute zero. Hence the 'critical line' lies along the horizontal axis ($T = 0$). The role of the correlation length is now taken by the mean spin-flip time τ for the localized spin. Thus one's task is to draw the lines of constant τ in parameter space.

When $\tilde{J} = 0$, the impurity spin is decoupled and one has the Curie susceptibility diverging like $\dfrac{1}{T}$ for small T. When $\tilde{J}$ is positive (ferromagnetic), it is clear on physical grounds that the coupling will simply tend to polarize the medium so as to increase the local magnetization; therefore one expects that the susceptibility will still diverge at low temperatures. By contrast, if $\tilde{J}$ is negative (antiferromagnetic), the coupling will tend to polarize the medium so as to compensate the impurity spin; then one expects qualitatively different behaviour, with the susceptibility remaining finite at zero temperature.

These simple arguments predict that the 'critical line' will be the semi-axis ($T = 0, \tilde{J} > 0$). Figure 13.2 shows a topology of trajectories consistent with the above. The point $\tilde{J} = 0$ is a fixed point; for $\tilde{J} > 0$ the behaviour reduces to that of a non-interacting impurity, and there is convergence to the fixed point $\tilde{J} = 0$; for $\tilde{J} < 0$, the behaviour is qualitatively different, and the trajectories move away to an as yet undiscovered destination, (a fixed point of type $\tau = 0$, contrasting with the $\tilde{J} = 0$ fixed point which is of type $\tau = \infty$). The topology of the trajectories near the fixed point $\tilde{J} = 0$ is typical of locally marginal situations, and recalls Figure 5.2.

Figure 13.2. Topology of trajectories on the $\tilde{J}$ axis for the Kondo problem

One sees how very simple arguments can elucidate plausibly the essential features of a situation in a way that is consistent with the results of calculation. Given this topological structure one predicts that the $(\tilde{J}, T)$ plane will divide into three regions: a trivial high-temperature region where the impurity is practically non-interacting, a low-temperature region where logarithmic corrections are important, and an antiferromagnetic low-temperature region governed by another fixed point (Figure 13.3). The change from one region to another occurs near the crossover temperature, denoted here by T_K, which because of the marginal nature of $\tilde{J}$ behaves asymptotically, for small $\tilde{J}$, like

$$T_K \sim \exp\left(\frac{-1}{|\tilde{J}|}\right) \tag{13.4}$$

176

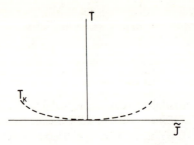

Figure 13.3. Sketch of the crossover
temperature T_K in the $(T, \tilde{J})$ plane

Before elaborating these results it is useful to consider the following anisotropic
generalization of the Hamiltonian (13.3):

$$\mathscr{H}_a = \sum_{k\sigma} \varepsilon_k n_{k\sigma} n_{k\sigma} - J_z S_z \cdot s_z(0) - J_\perp (S_x \cdot s_x(0) + S_y \cdot s_y(0)) \qquad (13.5)$$

Consider a $(\tilde{J}_z, g \equiv \tilde{J}_\perp^2 - \tilde{J}_z^2)$ diagram, with the temperature fixed at zero; the
isotropic case then corresponds to the horizontal axis. Notice that changing the
sign of $\tilde{J}_\perp$ is equivalent to a spin rotation, whence our choice of the parameter g. In
the plane $(\tilde{J}_z, g)$ there is a special line, a parabola, corresponding to $\tilde{J}_\perp = 0$, (Figure
13.4); the interior of the parabola is an unphysical region, corresponding to
imaginary values of $\tilde{J}_\perp$. On the parabola the problem is trivial because there is no
spin-flip for $\tilde{J}_\perp = 0$; from the renormalization-group point of view, the parabola is
a line of fixed points. The isotropy line ($g = 0$) touches the line of fixed points, and
by taking sections at constant g one can see that for $g < 0$ there are two fixed
points, which coalesce in the limit $g \to 0$. In this way one can once again understand
the local marginality of the isotropic case as due to the coalescence of two fixed
points.

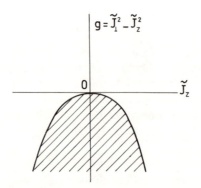

Figure 13.4. Regions of the para-
meter space $(g, \tilde{J}_z)$ for the anisotropic
Kondo problem

The renormalization group

We are now in a position to discuss how the transformations of the renormalization group apply to the Kondo problem, and to establish the differential renormalization equations. In the past the renormalization group has been applied in various ways stemming from its first or second versions, and the uniqueness of the results obtained by the different procedures has not yet been demonstrated in detail. Here we shall follow as closely as possible the same approach as to phase transitions in Chapter 4.

The first step is to integrate over those electronic levels having energy $|\varepsilon_k|$ between D and D/s, where D is the original cutoff parameter; (for instance, one can choose an electronic energy band that is symmetric, half-full, and of width $2D$). The second step is to change the cutoff parameter back to its original value by dilating the energies. In the present case no third step seems to be needed.

Under these transformations the Hamiltonian $\mathscr{H}$ transforms into a new Hamiltonian $\mathscr{H}'$; it remains to draw up the dictionary of correspondences. The 'primed' energies arise from the unprimed through multiplication by a factor s:

$$E \rightarrow E' = Es$$

In particular, we have for the free energy G and the temperature T,

$$G \rightarrow G' = Gs, \qquad T \rightarrow T' = Ts$$

An applied magnetic field H is renormalized similarly:

$$H \rightarrow H' = Hs$$

The calculation then consists in deriving differential renormalization equations for the parameter $\tilde{J}$. As usual, the first step introduces new coupling constants. For small $\tilde{J}$ one can express the new parameters in terms of $\tilde{J}$ through power series, thus obtaining a differential equation involving only $\tilde{J}$ itself, (i.e. an equation for the projection of trajectories onto the $\tilde{J}$ axis). One finds

$$\frac{d\tilde{J}}{dl} = C_2\tilde{J}^2 + C_3\tilde{J}^3 + C_4\tilde{J}^4 + C_5\tilde{J}^5 + \cdots \tag{13.6}$$

where $s = e^l$. Equation (13.6) determines the renormalization of the parameter $\tilde{J}$ when $\tilde{J}$ is small. One can check that $\tilde{J} = 0$ is a fixed point ($C_0 = 0$), and that $\tilde{J}$ is a marginal field with respect to this fixed point ($C_1 = 0$). The leading coefficients have been calculated; for any value of the spin S, one finds

$$C_2 = -1, \qquad C_3 = -\tfrac{1}{2}$$

From the dictionary of correspondences and the differential Equation (13.6) one can now derive homogeneity rules for the thermodynamic variables. For instance, the susceptibility $\chi(T, \tilde{J})$ can be written as

$$\chi(T, \tilde{J}) \sim e^l \cdot \chi(T \cdot e^l, \tilde{J}(l)) \tag{13.7}$$

Many results follow very simply from the homogeneous expression (13.7). Thus, keeping only the first, order $\tilde{J}^2$, term in (13.6), the differential equation integrates to

$$\tilde{J}(l) \sim \frac{1}{l + l_0}, \qquad l_0 \equiv \frac{1}{\tilde{J}_0}$$

Choosing the value of l so that $T \cdot e^l = 1$, one obtains for χ

$$\chi(T, \tilde{J}) \sim \frac{1}{T}\chi\left(1, \frac{\tilde{J}}{1 - \tilde{J}\log T}\right)$$

This approximate expression shows that in actual fact an expansion in powers of $\tilde{J}$ tends to turn into an expansion in powers of $\tilde{J}/(1 - \tilde{J}\log T)$, which accounts for the Kondo singularity at $T = T_K$ when $\tilde{J} < 0$. This apparent singularity, which has generated so much literature, is now seen to be an artefact that arises if the expansion (13.6) is cut short arbitrarily.

From (13.7) one can also estimate the susceptibility at zero temperature for $\tilde{J} < 0$. To this end one chooses a value $\bar{l}$ of l such that $\tilde{J}(\bar{l}) = -1$. Integrating (13.6) one finds

$$\bar{l} \sim -\frac{1}{\tilde{J}} + C_3 \log \tilde{J} + (C_3^2 + C_4)\tilde{J} + \cdots$$

whence

$$\chi(T = 0, \tilde{J}) \sim \exp\left(-\frac{1}{\tilde{J}}\right) \cdot (\tilde{J})^{C_3} \exp[(C_3^2 + C_4)\tilde{J} + \cdots] \tag{13.8}$$

The susceptibility, which at high temperatures varies like $1/T$ (Curie's law for free spins), saturates near the crossover temperature, so that one has

$$\chi(T = 0) \sim \frac{1}{T_k}$$

Accordingly, Equation (13.8) gives the variation of T_K as a function of $\tilde{J}$ when $\tilde{J}$ is small; one can see the provenance of the correction factors multiplying the asymptotic expression (13.4). The nature of the corrections follows directly from the non-singular nature of the expansion (13.6).

The expansion (13.6) is valid only for small $\tilde{J}$, and does not allow one to determine the destinations of the trajectories which for negative $\tilde{J}$ move outward from the origin. If nevertheless one keeps only the first two terms in (13.6), one can see that the equation identifying fixed points, $\dfrac{d\tilde{J}}{dl} = 0$, is obeyed at $\tilde{J} = -2$; this has led Abrikosov and Migdal (1970) to predict the existence of another fixed point at some finite distance. But in actual fact the value of $\tilde{J}$ obtained in this way is too high to allow the higher-order terms to be neglected, and the approximation is defective. The true result is that there are no fixed points at any finite distance. This is reasonable on physical grounds, (otherwise the singularities obtained would be too strong); it is almost proved by the analogy to the one-dimensional Ising problem with $1/r^2$ forces (Anderson, 1973); and it is confirmed numerically (Wilson, 1973).

Analogy to the one-dimensional Ising model with $1/r^2$ forces

For the anisotropic case (13.5) it is possible to derive a system of differential equations for $\tilde{J}_z$ and $g = \tilde{J}_\perp^2 - \tilde{J}_z^2$, in the limit where $\tilde{J}_z$ and g are small. In the physical region of the $(\tilde{J}_z, g)$ plane one then obtains the pattern of trajectories sketched in Figure 13.5.

The anisotropic case is interesting because one can establish an analogy between the anisotropic Kondo problem at zero temperature and the one-dimensional Ising model with long-range forces proportional to $1/r^2$. Thus a Kondo problem defined by the pair of parameters $(\tilde{J}_z, g)$ corresponds to an Ising problem at some definite temperature T_1, (the suffix I refers to the Ising model), and with long-range forces having some definite strength A_1. Under these conditions one can in the $(\tilde{J}_z, g)$ plane draw the physical line associated with an Ising model, obtainable by fixing A_1 and varying T_1. Such a physical line is sketched in Figure 13.6. The critical point of the Ising model corresponds to the intersection between the physical line and the horizontal axis. Several important features should be noted. If there did exist another fixed point at a finite distance $\tilde{J} < 0$, then this would imply the existence of a second critical point in the Ising model, which is highly unlikely; what is likely is that, for $T > T_c$, the trajectories move out to infinity, as is usual in phase transitions. For $T < T_c$, the trajectories move towards a line of fixed points, namely the bounding parabola. This behaviour entails some peculiarities. One can see that at T_c there is a discontinuity in the spontaneous magnetization of the Ising model, whence the transition is first order in Landau's sense, though second order if judged by the divergence of the correlation length; the discontinuity is compatible with the value $\eta_1 = 1$ (see Section 10.2) and $d_\phi^1 = 0$. Moreover there is a divergence in the

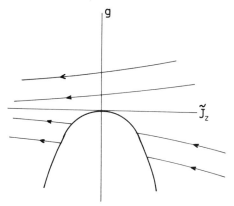

Figure 13.5. Pattern of trajectories in the parameter space $(g, \tilde{J}_z)$ for the anisotropic Kondo problem

180

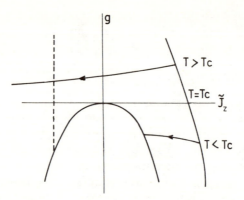

Figure 13.6. Same as Figure 13.5; the additional full line sketches a typical physical line for a corresponding Ising model. The broken line corresponds to a different model whose solution is known

exponent ν_1, governed by the anomalous dimension of the field g, (which is locally marginal in the neighbourhood of the fixed point at the origin).

The correspondence between Kondo problem and Ising model can be derived from a very general principle which is exploited also in transfer-matrix methods and in Feynman's path-integral formulation of quantum mechanics. In these cases one is usually faced by a correspondence between a problem in statistical mechanics (at finite temperature and in d dimensions), and a problem in quantum mechanics (at zero temperature and in $(d-1)$ dimensions). Most often the correspondence is exploited in order to solve the statistical problem by aid of the quantum problem (especially for one- or two-dimensional systems). Here one proceeds in the opposite direction. Another special circumstance should also be noted: from the fact that this Ising model involves long-range forces, it follows eventually that the equivalent quantum problem is a very special zero-dimensional one, dealing as it does with an impurity immersed in a Fermi sea of electrons; (the retardation of the interaction between the electrons, mediated by the impurity, is the reflection, in the Kondo problem, of the long range of the forces in the Ising problem).

Finally we mention that for a certain special value of $\tilde{J}_z$, $(\tilde{J}_z \sim -1)$, both these problems can be reduced to another soluble problem (Toulouse, 1969). Thus, the $(\tilde{J}_z, g)$ plane contains a line, the broken line in Figure 13.6, along which the solution is known. This circumstance is unusual; previously, the only known regions we have encountered have been the neighbourhoods of the fixed points, and a problem has been soluble if its trajectory came close to a fixed point. Here we have a new possibility which widens the region of applicability of the renormalization group, and deserves attention as a paradigm.

Throughout the discussions above we have considered only a very restricted parameter space; in actual fact parameter space is multidimensional and the true trajectories are more intricate than one might think from their projections as drawn in Figures 13.5 and 13.6. Here, as in the problem of phase transitions, this is a fact that should be borne in mind.

Low-temperature behaviour

The case where $\tilde{J} > 0$ presents no real difficulties and can be treated along the lines of Section 5.4. By contrast, the case where $\tilde{J} < 0$ constitutes an entirely novel problem. Qualitatively, one expects that at low temperatures the impurity will behave as if it were non-magnetic, because the trajectories converge to a fixed point corresponding to $\tilde{J} = -\infty$, and in this limit the impurity spin binds a conduction electron (when $S = \frac{1}{2}$) to form a singlet which is a non-magnetic entity. Thus one expects the susceptibility to remain finite, the specific heat to vary linearly with temperature, etc. The problem is to quantify these qualitative features. We shall concentrate particularly on the $T \rightarrow 0$ limit of the ratio $C_v/T.\chi$, where C_v and χ are the contributions of the impurity to the specific heat and the susceptibility. Considerations based on homogeneity suggest

$$\frac{C_v}{T} \sim \chi \sim \frac{1}{T_K}$$

and our task is to estimate this ratio as a function of $\tilde{J}$. We choose units such that for a strictly non-magnetic impurity

$$C_v/T\chi = 4$$

One approach is to evaluate the ratio for the special value of $\tilde{J}_z$ which allows an exact solution, (this gives $\dfrac{C_v}{T\chi} = 1$), and then to exploit the scaling properties; in the limit $\tilde{J} \rightarrow 0$ this yields:

$$C_v/T\chi = 2$$

In fact this value was originally obtained by Wilson (1973) in an ingenious numerical calculation, which constitutes a second approach, in a very different spirit, and consists in solving the problem for a small grain whose size is then increased progressively to infinity. A third approach (Nozières, 1974), yielding genuine theoretical insight into the result, is based on a description along the lines of the Landau theory of Fermi liquids, where one expresses the properties of the system in terms of just a few phenomenological parameters. This shows that the value $C_v/T\chi = 2$ is an immediate consequence of two physically plausible assumptions: first, that the perturbation is localized, and second, that the Kondo singularities are tied to the Fermi level.

The final outcome is that the ratio $C_v/T\chi$ varies with $\tilde{J}$ in the way sketched in Figure 13.7.

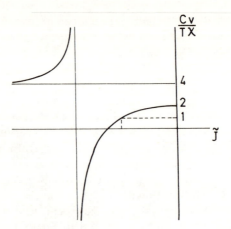

Figure 13.7. Sketch of $C_v/T\chi$ as a function
of the parameter $\tilde{J}$ for the antiferromagnetic
Kondo problem ($\tilde{J} < 0$)

Universality

The theoretical analysis of the Kondo effect has led us to define two simple
regimes, ($T \gg T_K$ and $T \ll T_K$), separated by an intermediate (crossover) region
which is complicated. This yields a very succinct qualitative description of the
situation. The theoretical analysis also allows one to calculate some numbers, as
we saw in the case of the leading coefficients in the expansions of T_K and $C_v/T\chi$ in
powers of $\tilde{J}$. There is no doubt that this quantitative analysis could be extended
systematically in the way suggested by the preceding chapters on phase
transitions. The task is to discover what properties are universal and therefore to
be calculated, and to obtain both the functional form of the scaling functions for
the various physical quantities (resistivity, specific heat, susceptibility, etc.), and
also the corrections to scaling. One should also study the effects of the different
physical perturbations that can act on real systems, but which are missing from the
idealized Kondo model (13.3).

13.4 One-dimensional metals

This is an enormous subject and we shall confine ourselves to pointing out some
landmarks for general orientation. For normal metals, the Landau theory of
Fermi liquids provides a phenomenological description of the low-temperature
properties. This theory is well confirmed in three-dimensional metals but has no
validity in one dimension; we could say that for Landau theory the dimensionality 1 plays the role of characteristic dimensionality, somewhat in the same way
as the dimensionality 4 plays this role for the Landau theory of critical phenomena.

Let $V(r)$ be the interaction potential between two electrons and $V(q)$ its Fourier
transform. When the range of forces is long enough for $V(2k_F)$ to be negligible, an

exact solution is possible, and one has a problem isomorphic to that of the infrared anomalies (Section 13.2); thus, the problem is characterized by a line of fixed points and by exponents varying continuously with a parameter (namely the coupling constant $V(q = 0)$).

If one includes interactions of arbitrary range, with $V(2k_F) \neq 0$, then the problem is no longer trivial. One notices then that the differential renormalization equations are isomorphic to those of the Kondo effect, at least in the lowest orders; the coupling constant $V(2k_F)$ now plays the role of $\tilde{J}$ in the Kondo problem. The repulsive case, $V(2k_F) > 0$, seems to be relatively simple, but the attractive case, $V(2k_F) < 0$, appears for the time being to be just as mysterious as the antiferromagnetic Kondo effect was originally. From this point of view the problem of one-dimensional metals seems to be the natural successor of the Kondo effect amongst the preoccupations of theorists, and it seems indicated to tackle it with the same prescriptions.

The aim is to discover how parameter space divides into regions each with definite characteristics, like normal or antiferromagnetic metal, superconductor, Peierls insulator, and possibly others. Of course in one dimension there can be no phase transitions at finite temperatures, but it may be that fluctuations of one definite type or another dominate at low temperatures. As regards experiments, there exist many metals which in a first approximation can be considered as assemblies of one-dimensional metallic chains, and which display remarkable properties.

Thus, the problem is of interest both to experiment and to theory, even though there is far to go from the simplest theoretical models to real systems affected by lattice type, by phonons, and by interactions between the chains.

13.5 Conclusion

We have learnt from this chapter, and should bear in mind, that the renormalization group can also explain exceptions from the rules; that phenomena attending marginality do occur in systems of physical interest; and that in the solution of such problems it can be useful to exploit structural analogies.

References

Abrikosov, A. A., Migdal, A. A. (1970), *J. Low Temp. Phys.*, **3**, 519.

Anderson, P. W. (1961), *Phys. Rev.*, **124**, 41.

Anderson, P. W. (1973), *Comments on Solid State Phys.*, **5**, 73.

Brézin, E., Zinn-Justin, J. (1976), *Phys. Rev. Letters*, **36**, 691.

Luttinger, J. M. (1963), *J. Math. Phys.*, **4**, 1154.

Nozières, P., de Dominicis, C. (1969), *Phys. Rev.*, **178**, 1097.

Nozières, P. (1974), *J. Low Temp. Phys.*, **17**, 31.

Thirring, W. (1958), *Ann. Phys.*, **3**, 91.

Tomonaga, S. (1950), *Progr. Theor. Phys.*, **5**, 544.

Toulouse, G. (1969), *Comptes Rendus Acad. Sci.*, **268**, 1200.

Wilson, K. (1973), in *Collective Properties of Physical Systems*, Nobel Symposium XXIV, Academic Press.

CHAPTER 14
Discussion and conclusion

'Leur oeil fixe m'attire au fond de l'infini.'   Victor Hugo

Every vigilant reader will have observed that this book contains passages that are clumsy or inconclusive. The responsibility rests partly on us and partly on the incomplete state of knowledge at the time of writing, as the reader will be able to judge for himself from the references we have given. In these circumstances, rather than attempt a summary, we shall end by trying to put into context what has been said; we shall try to do this partly by commenting on certain problems in field theory and elementary particle physics (Section 14.1), and partly by pointing out some directions in which the methods of the renormalization group can be developed, extended, or probed more deeply (Section 14.2).

14.1 Elementary-particle physics

Invariance properties and particle masses

To a particle of mass m is associated a length, namely its Compton wavelength

$$\lambda = \hbar/mc$$

where $\hbar$ is Planck's constant and c the speed of light. By virtue of the uncertainty relations, λ is the range of the interactions due to the exchange of such a particle. Thus, gravitational and electromagnetic forces, being due to the exchange of zero-mass gravitons and photons respectively, have infinite range and a $1/r$ power-law interaction potential.

In the analogy between the concepts of statistical mechanics and of particle theory, the correlation length ξ corresponds to the Compton wavelength λ, and the correlation function $\Gamma(k, \xi)$ corresponds to the particle propagator $\Gamma(p, m)$. Thus, the zero-mass case, corresponding to the critical point in statistical mechanics, is hallmarked by scale-invariance; this invariance is weakly broken if the mass is small, and one is then led to distinguish between the two cases $p \gg m$ and $p \ll m$, just as in statistical mechanics one distinguishes between $k\xi \gg 1$ and $k\xi \ll 1$. Eventually one arrives at the fairly general principle that scale invariance plays a role in high energy processes ($p \gg m$); in other words such processes, where the particle masses can be neglected, correspond to the critical regime.

The prediction, that at high energies one should observe a simple regime featuring power-law behaviour and homogeneity properties, has led to much experimental and theoretical work, especially on collisions of leptons (electrons, neutrinos, etc.) with hadrons (protons, etc.), and between protons.

Scale invariance should be compared to other symmetries that would be exact if certain masses were zero, but which are only approximate (i.e. valid only at high energies) because the masses are finite. Thus, the chiral symmetry of strong interactions would be exact if the pion mass were zero, but in fact holds only asymptotically for $p \gg m$ (pion). Notice that chirality is an internal symmetry, by contrast with dilatation which is a symmetry in space-time. Similarly, the local gauge-invariance of electrodynamics ensures that the photon mass is zero and that electromagnetic forces have infinite range. Continuing the argument along the same lines, it is currently thought that the spontaneous breaking of a local gauge invariance is responsible for the finite range of the weak interactions, and for the finite masses of the charged and neutral bosons which mediate it.

Asymptotic freedom and anomalous dimensions

The propagator of a free particle with mass m_0 is

$$\Gamma_0(p, m_0) = \frac{1}{p^2 + m_0^2}$$

(compare this to the Gaussian correlation function in statistical mechanics). Hence, for $p \gg m_0$,

$$\Gamma_0 \sim \frac{1}{p^2}$$

But in the presence of interactions, (compare again with statistical mechanics), one has for $p \gg m$ (i.e. in the critical regime),

$$\Gamma \sim \frac{1}{p^{2-\eta}}$$

Accordingly, one can say that in this regime the particle behaves as if free (this is the idea of asymptotic freedom), up to a change in the exponent (this is the idea of an anomalous dimension). The exponent is universal, i.e. independent of the coupling constants, over a wide class, except for marginal cases.

The concept of asymptotic freedom underlies models like the parton model, where it is assumed that at high enough energies the constituents (partons) of say the nucleon behave like free particles. What is new here is the idea, clarified by statistical mechanics, that the presence of anomalous dimensions will serve to distinguish the regime of asymptotic freedom from a regime totally void of interactions.

Bare and physical coupling constants and renormalization

Another idea important both in elementary particle physics and in the many-body problem is the renormalization of interactions. Consider for instance the Kondo problem (Section 13.3) for some given negative value of the coupling constant J. It is then convenient to define an 'effective', 'physical', or 'renormalized' coupling constant which is a function of energy (or of temperature). At high energies the renormalized coupling constant J_R assumes the bare value J; but as the energy decreases, the coupling constant decreases, tending to $-\infty$ as the energy tends to zero. The general physical idea is that at low energies a particle is 'dressed', i.e. surrounded by a polarization cloud, so that the 'quasi-particle' consisting of particle plus cloud reacts as a single entity; by contrast, at higher energy the cloud is less and less able to follow the particle, whence at high enough energies the latter alone reacts, behaving as a 'bare' particle. This naturally suggests that the renormalization group should be able to describe the crossover, i.e. should be able to interpolate, between the 'bare' and the 'physical' limiting values of the coupling constants; in particular one is led to hope that in this way it will become possible to account for the observed values of certain coupling constants (like the fine-structure constant $\alpha = e^2/\hbar c$ in quantum electrodynamics).

14.2 Outlook

In this final section we aim to indicate some issues that could well be pursued in greater depth, detail, or generality. We must also stress certain features of the treatment in earlier chapters that remain unsatisfactory.

The great variety of ways in which the renormalization group can be applied raises the twin problems of uniqueness and of optimization; criteria should be found for obtaining the best results from approximate calculations, for the best choice of lattice or continuum model, and so on.

For $d > 4$ and the case of an ordinary critical point, (in general for $d > d_c$), the dominant terms are those given by Landau theory. Nevertheless several problems remain unresolved, for instance the determination of the subdominant singularities, the topological description of tricritical points and of first-order phase transitions, the significance of unstable fixed points, and many others.

It would be desirable to have a systematic way of describing the low-temperature ordered phase $(T < T_c)$ in terms of parameter space and of trajectories, and also to elucidate further the effects of angle variables on the low-temperature phase and on the equation of state, not only when $n > 1$ but also when $n < 1$.

To date, the possibility has not been excluded that complex eigenvalues may be found in the spectra of fixed points. Though this situation has never yet been encountered in the problems studied in statistical mechanics, it is common in the theory of dynamical systems, and can give rise to limit cycles and other phenomena. It is quite unclear what such phenomena could correspond to if they do exist in statistical mechanics.

More interesting is the case of complex fixed points. We have already encountered systems where two real fixed points meet and then disappear, giving rise to a pair of complex fixed points. In some such cases one would expect to see 'evanescent scaling', with asymptotic behaviour given by cut-off power laws. This amounts to saying that the correlation length becomes large though it fails to diverge, (it becomes the larger the smaller the imaginary parts of the coordinates of the fixed point). This concept should prove useful in many systems.

Dynamic effects, deliberately ignored in this book, also rate a mention; for them, the existence of homogeneity properties and of universality remain to be elucidated.

A whole series of desirable but difficult extensions include quantized percolation and the problem of fully-developed turbulence.

Finally we refer back to the body of the book where we have mentioned many open problems which it would be pedantic to recapitulate here.

In a wider context there arises the question of the extent to which the language and the concepts of the renormalization group may prove useful in less closely related disciplines. To illustrate the generality of this language we mention the work of Jona–Lasinio on probability theory, which sheds new light on the concepts of the central limit theorem, of stable distributions, and of catchment areas.

Index